Seasons of
Heron Pond

Seasons of Heron Pond

by Mary Leister

drawings by Charles Hazard

Stemmer House
PUBLISHERS, INC.

Owings Mills, Maryland

Inquiries should be directed to:

STEMMER HOUSE PUBLISHERS, INC.
2627 Caves Road Owings Mills,
Maryland, 21117

Published simultaneously in Canada by Houghton Mifflin Company Ltd., Markham, Ontario

A Barbara Holdridge book
Printed and bound in the United States of America
First Edition

Library of Congress Cataloging in Publication Data

Leister, Mary.
    Seasons of Heron Pond.

    "A Barbara Holdridge book."
    1. Pond ecology. I. Hazard, Charles, 1947–
II. Title. III. Title: Heron Pond.
QH541.5.P63L44 1981   574.5′26322   81-9408
ISBN 0-916144-84-4   AACR2

Contents

Foreword

"I have traveled a good deal in Concord," wrote Henry David Thoreau in *Walden*—omitting to mention that he had also traveled a good deal elsewhere—but he traveled mostly in Concord, and he wrote with authority of Concord's marshes and ponds and meadows and woods because he was interested and involved and so thoroughly there.

Most writer-naturalists have a Concord of their own, one corner of the world, be it as small as their own home farm or as large as the Gobi Desert, where the happenings and the inhabitants and the terrain are so dearly familiar and understood as to be almost ingrained; and so important, to them, in the livingness of the whole earth that they must translate it all and send it forth so that others, too, might love and understand.

When *Wildlings* was first published, several years ago, I felt somewhat defensive about the "one square mile of Maryland countryside" of which I wrote. Today I realize what a glorious privilege has been mine, to have spent thousands of joyously uncounted hours ranging over the very same fields and hills and valleys, wading the very same marshes and streams and mudpuddles, poking into the very same nooks and ledges and fence corners, and watching over the very same pond for seventeen long and adventuresome years.

But nothing is, of course, wholly the same. The very passing of so many years brings inevitable changes to the land and to the walkers upon it.

Both Kon-Tiki and Kela, my Great Pyrenees friends of *Wildlings* fame, now run as happy shadow-dogs beside Kuonny and me as we wander the wilderness country they loved. Kon-Tiki, adventurous spirit, first led me into all of this new territory, and she taught me how to learn its secrets. Kela, because I was braver, pressed the boundaries a little wider, familiarized me with every hidden corner, and reinforced, in her own way, all the lessons Tiki first had taught.

Kuonny is not a Great Pyrenees. He is a dog of Very Special Bloods, admixture unknown, a sizeable fellow in chestnut-and-white. He is my friend and overseer who insists, as did his predecessors, that we go out into the wild land in every weather to trot, to run, to walk on the double—but never, never to stroll—and he is thus responsible for my continuing very excellent health and great spirits.

But Kuonny also "stays" just perfectly, and so he is responsible for the fact that I still know the large and the smaller denizens of this land in somewhat close detail, that I know who lives and who dies, who comes and who goes, and what changes are wrought on the land and its inhabitants by storms, by droughts and by human stewardship.

That many of these changes are serious and probably irreversible I must acknowledge. Heron Pond, once twelve feet deep and covering more than two acres of pasture land, has lost at least one-third of its area and more than one-half of its depth. The marsh below it, a watery area of reeds and sedges and deep, black, boot-sucking mud, now grows goldenrod and rough-stemmed asters and is traversible, with boots, in almost any season.

My "one square mile" of free-wandering territory, once comprised of four woods, three marshes, three creeks, two pastures, two ponds and several very different fallow fields, has now shrunken, for varied reasons, to probably less than one square half-mile. But that square half-mile holds the very heart of my old territory and I walk its mainly linear ways with joy and sincerest gratitude.

I still have one pond, one stream, one marsh, one wood, one fallow field, several hills, and a lovely valley; and if Emily Dickinson's one clover and one bee made, for her, a prairie, this one-half mile of countryside with its seriously depleted but still tenuously viable wild community makes a whole world for me.

Not, I hope, that I am parochial about it, for I have wandered far from my Concord. I walk this narrow wilderness with senses and intuitions honed by having walked wild paths in other lands and in all but four of our contiguous forty-eight states. I have met exciting and exotic animals and plants, and I hold in my heart many distant and dearly loved places, but this land I walk day after day is sacred ground; this land is Home.

What happens to this small section of the living earth and to the life it nurtures is the daily interest and the continuing adventure of my life. I walk its wild, wet ways in peace and contentment. I walk with zest and in the best of health. I could not ask for more.

Mary Leister

Seasons of Heron Pond

Unwilling Harbingers
of Spring

$\mathcal{A}$t the foot of a ravine in the middle of the woods, a curving pool of backwater eddies gently against the eastern bank of the northward flowing stream. The climbing suns of early March shine through the leafless limbs of the creekside trees and warm the quiet waters of the pool, the clay shelves that rise behind it, and the bits of sticks and stems and sodden leaves that lie adrift on its surface.

I watch this pool, these early days, because the warmth of the sun on its sheltered cup is sure to bring a melancholic group of water striders from hibernation quarters under the detritus on the bank, or from the mud beneath the faintly turning waters.

Try as hard as I can, I have never seen them emerging. I always catch the droopy vanguard already out and looking as though they wish they hadn't come. They stand about on the water skin with neither verve nor interest, and if the small current swings one of them against another they both extricate themselves from the leggy tangle and stump off across the water to stand listless and dull in a new location.

These striders, I think, are forced from their winter quarters by the heat of the sun on the dark surface of the bank, but when they move out onto the open water there is not enough warmth to stir them to further activity. So they stand about in murky consciousness until the sun slides away behind the woods and the lowering temperature allows them to creep back into their dark caves again. But they *do* come out—too early—and I rejoice to see them.

I have loved water striders since I was five years old and walking "up the hollow" to the one-room school at the head of the creek. My path was a cow path that followed the winding of the stream among alder bushes and willow trees, through patches of skunk cabbage and gardens of trout lilies, and, at almost every turning of the way, revealing small tribes of these magical creatures who walk with grace upon the waters.

I remember so clearly the afternoon I discovered the water skin on which they moved. I had parted from my school mates at the bridge (loose planks laid across logs), where I crossed and they didn't, and, alone, I lay down upon the bridge with my head hanging over to watch a crowd of water striders at their wondrous skating.

Some chance angle of light showed me the round dimples their hair-thin feet made in the top of the water. I could see the stretching of the water surface beneath each foot like a pressing-in on the thin rubber of a transparent balloon. There came a silence, a dizziness, a glimpsing of a world beyond my reaching. I can still feel the rainbow-colored wonder, the yearning, the need to part the curtains. . . .

Although I could plainly see the skin in the dimples where the striders stood, I couldn't see it anywhere else—nor feel it. My hand, laid upon the water lightly, so lightly, could not lie atop it. Even my tiniest finger went through that elastic skin without the slightest sensing of its texture. And my feet, in laced leather shoes, did not do better. . . .

Today, I stand on the banks of a small open pond in a Maryland pasture. The sun has been warming these waters for several weeks and the water striders skate upon the surface nimbly alert, wide awake, and moving into their warm-weather life.

Their long brown bodies look fresh and newly stitched with white. They glide so easily on widely parted legs. The long rear legs do the balancing and steering. The middle legs are the oars. They are equipped with tufts of hairs for paddles, and it is their smooth, sweeping motion that scuds the strider over the water. The front legs, although they extend well beyond the head, are shorter than the others. They provide front-end balance as the strider stands or skates about, but they are also vises which capture and hold small insect prey, like the front legs of a mantis.

Even as I watch, one strider leaps upward, into a swarm of gnats five or six inches above the water, and comes down with an almost invisible captive clamped in the angle of its folded forward legs.

The strider inserts its sharp beak into the gray body, drains its life juices, and casts away the chitin shell. Another strider dips suddenly forward and grasps in its front claws some creature I cannot see. I cannot see, either, whether it is pulling its captive from beneath the water skin or merely lifting it from the surface.

There is mating activity, too, upon the sunlit waters. In the heady warmth of this springtime afternoon an amorous half-dozen males, with neither tact nor finesse, are bluntly mounting or attempting to mount the scattered females of their tribe. Legs wave in the air as overzealous males are roughly rejected. Annoyed females, their bodies held nearly vertical, their rears down, hop unattractively away. There are leap-frogging pursuits through the midst of the skaters and half the bystanders fall victim to head-bangings, leg-tanglings, or watery up-endings. But all the hubbub is not suffered in vain. Occasional females stand blandly receptive.

Their eggs will be laid all through the warm spring and summer months. The female striders will pierce the water skin with their ovipositors and place their eggs in long, parallel rows, just under the water surface, on the vegetable debris that floats in the water or on the edges of the bank itself.

The miniscule creatures that hatch from these eggs will be nymphs that look very much like adult water striders. They will skate on the water's surface, eat microscopic animals, change into new skins and grow larger and larger until, with their final molt, they become adult water striders.

Some of them will have wings, but not all, and those who have them will seldom fly. All of them have oil glands in their feet which help to keep them on top of the water, but they can also go under water if they wish. All are covered with tiny hairs, like velvet overalls, and they can slip beneath the water skin with silvery bubbles of air caught all about their bodies.

A small part of this pool is shaded from the sun at this hour, and the striders are concentrated in the sunlit portion. I am watching their activities upon the surface, but I am also watching the shadows thrown on the clay bottom by the dimples their feet are making in the water skin. The shadows from their feet I expect, and I see. The shadows from the dimples I must often have seen before, but, suddenly, this afternoon, I am aware of them. Transparent curves in a transparent substance—and shadows are thrown on a clay stream-bed.

But there is more.

Small intermittent breezes blow about me, so slight as to be scarcely felt, so slight as not to ripple the water, not even faintly, to my sight; but I clearly see on the bottom of the pond the curling shadows of their passing.

The Barred-Owl Caper

In the darkness of an April night I awakened to the hooting of a barred owl so close by that it seemed to come from inside the house.

As I lay in bed listening, the owl obligingly hooted again, and the sound rang so loudly through the house, I was sure, because all the windows were flung open to the almost-summer warmth of the night. Then, in the distance, from down among the pines, I heard the owl's mate call a long eight-hooted reply.

"All's well with the world," I thought drowsily and drifted back to sleep—only to be brought sharply and thoroughly awake by a noise, certainly inside the house, that sounded to me like a handful of yardsticks clattering lightly down a wall.

I was out of bed in an instant, not so much alarmed, I think, as I was curious. The two cats and the dog were sitting up, their heads cocked, looking surprised and interested but not frightened or ready to do battle. They looked very wise and knowing, I thought, as though they knew what had happened and that it was really nothing to be concerned about. But I had to investigate.

My family was away for the weekend, my animals were all in sight, and the only yardstick I possess was standing quietly in its proper corner in the study closet. I could find nothing amiss, nothing out of place either upstairs or downstairs, so I crept back to bed and, in spite of the continued and very loud hooting, went back to sleep.

My dog awakened me, asking to go outside, just as first light was breaking in the east. I stood for a moment in the open doorway, enjoying the spring morning, and the barred owl hooted again from almost directly overhead. I stepped out into the dooryard, looked up to the roof of the house, and the owl hooted once more—from *inside* the chimney. "Oh, no," I wailed and went in to make sure.

It was not in the chimney of the livingroom fireplace. No way to look up the furnace chimney. I opened the damper of the basement fireplace and a handful of last year's dried peony stems clattered through onto the grate—like so many yardsticks clattering down a wall.

I peered up the chimney. There, down inside but very close to the top, was the barred owl, its back pressed against the flue, its wings spread round nearly full circle, clinging to the smooth ceramic lining of the chimney, and its feet, still holding a few dried peony stems, braced widely apart on opposite sides of the chimney lining. The owl peered warily down at me, made the slightest of upward movements—and slid downward another five inches.

I left the damper open so there would be a circulation of fresh air for the owl to breathe, and tried to think how I could set it free. Well, I couldn't. I couldn't even handle the ladder—and that put an end to any speculation on the second step of a rescue.

The owl in the chimney, having called all the night, still called at least once every five minutes, and its mate in the pines called back; but the free owl did not fly up to keep the imprisoned owl company or even to have a first-hand look. Their ceaseless exchange of calls was apparently enough. I wondered, more than casually, if their calls were a simple, repetitive, "I can't come home," "Why not?", or an explicit question-and-answer description of what was happening, or only an owl-type of covey-call, a keeping-in-touch.

I phoned everybody I knew who was interested in birds, and a great many people I didn't know, and I drew a blank. It was a holiday weekend and everyone was out of town or otherwise unreachable. And hour by hour the owl kept slipping down the chimney, away from that small, round opening toward the sky.

Sometime in the early afternoon, with a brushing slide and a clattering down of the remaining dried stems from its talons, the exhausted owl dropped the last dozen feet and stood with relaxed and folded wings upon the cast-iron grille work over which the metal vanes of the damper close.

If the owl had fallen into the living room chimney where the closure is a simple folding-down of two metal plates, it would have dropped into the fireplace and been free inside the house—

which would have been another problem, but a problem more my size.

But this damper, for whatever reason, is protected by a metal grille, and the grille is cemented into place, and, while its open spaces are nearly twelve inches long, they are scarcely two inches wide. Perhaps the owl could have been pulled through that two-inch space. I don't know. It looked so large standing there in the semi-darkness. But so much of an owl's size is the soft spread of its feathers. . . .

Down there, at the bottom of the chimney, the imprisoned owl no longer hooted. But its mate, never moving from the nest area in the pines, sent out quavering, questioning-sounding calls through the long, hot afternoon.

About four o'clock of that same afternoon my brother arrived and the rescue operation began. Up against the roof went the ladder. Down the chimney went an empty grain sack tied at its four corners with four lengths of clothes rope and weighted in its center with a few small stones.

Having both witnessed and read about the ferocity of trapped owls, my brother was prepared for violence with this one. His hands protected by heavily lined, heavy-duty leather gloves, he reached through the narrow grille and urged, cajoled, pushed and prodded the owl into the center of the flattened sack. And the owl quietly obeyed, as though it understood, without a single snap of its fearsome beak and without the slightest attempt to fight those hands away.

While I sat on the basement floor and talked to the owl, my brother went back to the housetop and carefully pulled on the ropes. The sack rose four or five inches and the owl, without moving its feet or flapping its wings, rolled helplessly off the sack and down onto the grille again. (We could not put the owl inside the sack because we could not reach up through the grille far enough to manipulate both bird and sack.)

Another try, and still another. Then a few more stones in the middle of the sack to make its concavity deeper. But always the owl, not fighting in any way, just simply losing its balance, rolled out of its rescue hammock.

Now my brother, reaching gingerly through the grille openings, tied the owl's feet together. The owl stood quietly, watching every move with its big brown eyes. It submitted, without a gesture of defiance, to the bonds about its ankles, and it flopped over obediently when sack and bird were brought together.

Back to the housetop went my brother. Carefully, so carefully, he pulled on the ropes and sack and bird began the slow ascent up the flue. And now there was the fear that the owl would escape at the top of the chimney, fly away with its feet tied together, and be in as bad a situation as when captive in the dark tower.

But the owl remained motionless, cradled in the burlap sack a few inches from freedom, while my brother reached down into the chimney opening and cut the bindings from its ankles. Then he lifted the owl out and set it gently on the stone rim of the shaft.

The owl, its feathers all awry and darkened with the dust and soot of a thousand wood-fires, blinked in the light of the low-hanging sun, then turned its head and looked at my brother standing there beside him on the housetop.

The look was such a long and studied one that my brother reached out his hand and brushed the top of the owl's head. The owl blinked its eyes and sat on, still gazing fixedly at my brother who, by then, was petting it as though it were a small dog.

After what seemed a long while, the freed owl spread its wings and took off in a graceful arc over the housetop and down to a great oak tree near the creek where it perched well out on a limb and began a careful preening of its feathers.

Strangely, during the hour, plus, of the actual rescue operation, the owl in the pines was silent. But when the freed owl flew to the oak, its mate immediately resumed its questioning calls.

For a full half hour the once-captive owl preened its feathers and did not answer. But when it finally did reply it gave a long series of hootings and gutterals, paused for the mate's reply, then repeated the same series of hoots and noises in its throat. Then it fell silent, went on with its preening, while its mate continued its anxious-sounding calls.

Just before true darkness fell, the freed owl lifted itself into the air and flew like a great dark moth down to the pines. I listened intently, expecting to hear a long dialogue of hootings and chucklings and gutteral moanings, but I heard not a sound from the grove of the pines.

A Ladder for an Evening One

I stood alone, one late spring evening, on the uncertain floor of an abandoned power house on the edge of a swift-flowing river.

The sun had already set. Deep shadows were creeping around me from the dusty, cob-webbed corners of the old, wooden building. Wheels and pulleys, shafts and belts, dismantled engines, broken tools, piles of wood and heaps of iron cluttered the floor. The place smelled of rotting wood, of rusting iron, of spilled engine oil, and of mice.

A crude six-foot ladder, with two of its rungs in place and the others hanging askew from one or the other of its standards, leaned against the wall. The tops of its standards extended an inch or two into the opening of a gaping square, just higher than my head, where a window once had been. The seven-fingered leaf of an elderberry clump waved in at the opening, and through it, against the dull rose glow of the sky, I could see the erratic rushings and swoopings of a colony of dark-winged bats on their early evening hunt.

As I stood there in the murky silence, watching the bats and listening to the faint settlings of the ancient wood, a sudden shrill and piercing cry from the jumbled debris at my feet stiffened my ear lobes, tugged at the roots of my hair, and sent a shiver up my spine.

The cry came again, instantly, constantly and furiously; and my startled reactions subsided before the very numbers of the cries. But I couldn't locate the source of all that clamor.

I searched over the piles of rubble, muddled and indistinct in the increasing duskiness, but I could find nothing. Moreover, the shrilling shifted from one location to another with amazing rapidity. Either there were several somethings screaming their woes or the one with the troubles was an acrobat.

I fumbled my flashlight from my knapsack and examined

each piece of trash, each pile of rubble. Nothing. But the incessant ringing cries continued until they echoed in my ears.

I turned my attention to the sparse areas of open floor between the wasted discards. Just dust and rust and scraps of wood, bolts, and washers, and rusted nails were there—and, then, one flick of motion in the midst of it all. One tiny, ungainly bit of dust-covered fur slowly humped itself up and flopped itself down and advanced a quarter of an inch across the floor in front of me while it uttered the loudest and most indignant protests I have ever heard.

It was a baby bat. A tiny Evening One.

He was scarcely more than half an inch long from the point of his irate nose to the tip of his embryonic tail. His body was clad in velvety brown fur, and his eyes were screwed down in the wrinkles of his tiny ugly face. His upright ears looked too big for his head, and his soft skin wings were only a cumbersome burden to be dragged beside him through the dust.

But where had he come from? There were no other bats in the room. He was the only one, making all the shrill cries by himself. The ventriloquistic effect came when he turned his head or when his voice met the baffle of some piece of debris on the floor.

With my light I searched the ceiling high above and found just one small opening where the end of a narrow board had rotted through. Standing beneath that opening I could hear the muted squeakings and cryings of unknown numbers of other baby bats whose home roost was obviously the loft above.

And now I knew the story.

This little bat, and the bats flying outside the powerhouse in the semi-darkness, belonged to the genus of little brown bats, the most common, most numerous of the bats in our country.

From the time a little brown bat baby is born until it is two weeks old it clings to the fur of its mother's stomach with its wing claws and the claws of its hind feet. Head-downward, because its mother is hanging head-downward, it feeds from her small nipples. Head-downward, head hanging back between her legs, it goes out with her on her nightly flights.

But when the baby bat is two weeks old it has become too heavy for the little mother to carry as she flies about to feed, so she leaves it behind for the night, safe in the home roost with all the other two-week-old babies.

The baby bats' eyes now open, and, aware of one another for the first time, they begin to crawl about, to touch, and soon to chase one another about. And many a little bat loses its hold and goes falling through the air to the floor or to the earth below.

The frightened baby flaps his wings frantically on the way down, and probably breaks his fall that way, but all he can do when the fall is over is to crawl slowly and protestingly all the way back up to his original perch.

Until, of course, one night he discovers that he can fly. And a whole new world opens up before him.

This shrieking baby bat on the floor beside me had possibly fallen directly through that hole in the ceiling; or perhaps he had toppled through as he crept toward a wall to climb. His eyes looked as though they had only just opened. This might very likely have been the first night he had been left behind by his mother. And he spoke vociferously of his discomfort and aloneness.

What could I do for this baby bat? I could not reach that single hole in the ceiling where it had fallen through. I found the stairs to the loft but they were decayed and fallen. In any case, the door at the top was sealed shut by heavy planks spiked across its face. So, since I could not return the baby bat to its home, I decided to stay around and see how the baby would handle his own predicament.

What he was doing at the moment was to crawl in a straight line across that open space of floor. He humped himself along on the bony elbows of his wings and his curved-clawed feet. He flopped his impossibly tiny half-opened wings at every quarter-inch step, and every time he flopped he uttered another angry and insistent cry.

I have no idea how much time he spent creeping across that floor. Full darkness had fallen, I know, and I watched him in the beam from my light which did not seem either to annoy or to attract him in any way.

Presently his round pug nose touched against the nearest standard of the ladder which reached to the paneless window above us. Here, at last, was something vertical, and the baby bat, still screeching, immediately set himself to climb it.

He caught the infinitesimal thumb-hook of his left wing into the rough wooden surface and drew himself flat against the standard. Then he stretched his other wing above his head, caught that thumb-hook into the soft old wood, and pulled himself up—scarcely one-eighth of an inch.

Four such repeated moves and he was hanging on the ladder standard, his little clawed feet just barely off the floor. And thus he climbed, hand over hand, or wing over wing, literally pulling himself up the ladder, and never ceasing for one instant his thin insistent shrilling.

It took that baby bat considerably more than an hour to climb the ladder, and, when he had done so, he sat on top of the standard in an ungainly hump, like a miniature gargoyle, and howled his protests to the immensity of the earth and the night-dark sky. But he did not howl for long.

I saw, for one fleeting instant, in the beam of my light, a fluttering of dark bat wings at the top of the ladder—and the baby bat was gone.

I can never *know* what actually happened, but I think the mother bat heard her baby crying in the window, that she picked him up, and carried him off with her into the night.

A Three-Bullfrog Afternoon

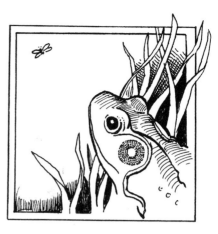

Four o'clock on a June afternoon. I sit on the grassy curve of the earth dam, close to the water, watching the ceaseless, hovering, up-and-down activity of several tiny, pale-lavender damselflies. They are placing their numberless miniscule eggs, one at a time, on the undersides of the leaves of the duckweed rafts that float in the quiet shadows under the pier.

The sunlight is tempered by a golden haze, but it sparkles on the lazy ripples of the pond and lies warm and somnolent on the willows. Lush June grasses have crept to the very edge of the water on the low sloping banks of the eastward shore, where dragonflies zoom and crayfish build their castles and sleek brown muskrats swim ashore to graze.

Three male bullfrogs live over there on the eastern side of the pond, each one feeding in his own sector of the waters, each one, now, calling from his own special niche among the matted roots and grasses. Since it is a big pond and there are only three bull-frogs, each one is separated by more than a hundred feet of pond-side from his equally exclusive neighbor.

This is bullfrog mating season, this summery month of June; but not for them the gathering in cooperative choirs and choruses, the barbershop harmonizing, the love-blind hobnobbing with the froggy proletariat that so characterizes the spring rituals of their lesser relatives. Each of these bullfrogs sings his own lovesong all by himself, he attracts his own lady or succession of ladies by his own efforts, and he woos and wins her or them in the privacy of his own watery holdings.

I had a close look at each of these bullfrogs a week ago when they had only recently pulled themselves up from the pond bot-tom, when they were not yet recovered from their winter-through-spring brush with death. Their reactions, last week, were slow and sluggish and they were certainly not singing love songs.

It seemed to me that it was not so much that their muscles were stiff from long disuse or that their unadjusted eyes were not seeing me near, but that there was simply not enough bullfrog consciousness on tap to order a leap to safety.

Even at their best, though, bullfrogs don't look especially bright. They are big when they reach maturity, and these were seven or eight inches long, plus a generous three inches wide at the rear; but their eyes are surprisingly small, their heads are broad and flattened, and they sit with their heavy bodies in a flat-crouched position that looks altogether bull-like—single-minded, pugnacious and dull.

But these three frogs, with their winter weariness still upon them, were actually more dull-witted than they looked.

They had all been sitting out on their basking-shelves in the bright, warm light of day for some time, and the skin of each one had changed from the near-black they wear when lying in the mud or living in deep water, to a bright olive-green alive with yellow color, and spotted a very little with brown. Even their throats wore a hint of yellow, an indication that all three were males, even though, this far south, both male and female usually have throats of dirty-white, sometimes mottled with brown.

The fact that all three bullfrogs are calling, on this mating-season afternoon, does not in itself establish their maleness. I have never witnessed this, but I have been told that the female bullfrog can croak, too, and in the same unmistakeable style as the male. She just doesn't croak as forcefully—and for a good reason. She has only one bubbling throat-pouch from which to develop her sound, while he bulges out a long pouch on either side of his head, between his ear and his arm, and puffs up a third vocal sac in his throat as well.

But not one of these frogs was in condition for a basso profundo performance last week, so it was neither their voices nor their vocal sacs that declared their maleness to me. It was the size of their ears!

Behind a bullfrog's eye, and below it, in the curve of the skinfold that runs from his eye to his arm, lies a noticeable circle. This is the frog's ear. It is simply an eardrum covered by the membrane of his skin and it is called the tympanum. The tympanum of the male is much larger than his eye. The tympanum of the female is small, about the same size as her eye.

So the tympani of these three frogs are far larger than their eyes, each one wears a hint of yellow on its throat, and they roar their to-whom-it-may-concern lovesongs across the pond, with all the force of three vocal pouches apiece. I say the three are *male* bullfrogs.

Usually, without any particular thought on the matter, my ears transliterate bullfrog songs into the traditional "Chug-a-rum" or "Jug-o-rum!" or "Better-go-round!" But this afternoon all I can hear in these three well-separated calls is a sound identical to the far-off bellowing of a very angry Holstein bull. The sound is unmistakable. Only a bull bellows like that.

Yet, as a bull's bellow, it sounds very far away. The nearest bullfrog is squatting in the shallows, its head lifted just above the duckweed, scarcely 300 feet from where I sit; and his frog-call, loud and clear, comes from there. But when I hear it as a bovine roar it sounds as though it were several pastures and a low hill removed from the pasture in which our jointly shared pond is located. Yet, distant or not, it is definitely a bull's bellow, and now I know, at first hand, the very likely reason why this frog is known as a "bull" frog.

But the concert of bull's bellows is suddenly cut short. My neighbor's little dog leaps the inlet stream and comes trotting up the eastern shore. A green frog on the bank leaps into the air and comes down on the water with a mighty splash that sends a warning sound around the pond and rouses quick circles of ripples to hide his own descent into the muddy bottom—and to warn those creatures under water of the approach of unknown danger.

The two bullfrogs out on the bank slip into the water without a sound, without a stirring of the grasses, without the faintest splash. The third one, already in the water, disappears into its depths with only one detectable ripple.

For bullfrogs, though they croak unceasingly for hours at a time, or sit as silent and unmoving as moss-covered stones the whole morning through, are, nevertheless, alert in every nerve with every muscle cell at the ready for instant action a whole moment *before* it is required.

A speeding dragonfly zooms by a bullfrog and the dragonfly is no longer there to be seen. A small bird hops down the bank for a drink at the water's edge. The bullfrog does not move but the

bird disappears—except, perhaps, for a few tailfeathers still protruding from the frog's green jaws.

A great blue heron fishing in the shadows of a willow stabs
his pointed bill, but the frog is three feet away under a mat of grass
when the bill reaches the spot where the frog was sitting. A man,
a frog-hunter, slaps his net over the bullfrog on the bank and all
that he finds under it is grass and muddy water. But there are also
hawks and owls and otters and foxes and dogs. . . .

In the water, the always-hungry bullfrog skims water striders
and back swimmers, whirligigs and mayflies from the pond's surface; gathers crayfish and snails, water beetles and dragonfly larvae
from the bottom; gobbles fish and smaller frogs, tadpoles and little
turtles up and down its waters. But in those same waters there are
otters and water snakes and full-grown snapping turtles and great
blue herons. And man, as a hunter of frogs. . . .

On the bank or in the water, the bullfrog is nearly invisible.
He moves without stir and with exceeding rapidity. He moves in
absolute silence. But so do those who would eat him.

Moreover, the bullfrog has never learned to conceal himself
properly when he dives into the mud. He buries his head, his
shoulders, his arms and his legs; but the great, wide girth of the
rest of him is often left unhidden—a dark protruding lump on the
top of the mud, highly visible to a great blue heron, an otter, a
turtle, a snake, or a man who is hunting frogs. . . .

Bumbling Bumblebees

The August sun was up. Its light flowed down the lush valleys on either side of the house and gilded the green hills beyond, but the building and its gardens still lay in deep shadow, curtained from the morning by the dark-leafed woods rising to the east.

In the cool protection of this shadow a drenching dew hung heavily on all the summer green, silvering the clematis vines on the fence posts and dripping from the tips of leaves in the shrubbery. Neglected lawn grasses curved low to the ground, weighted down by teardrops of crystal that showered the earth about my booted footsteps. Every flower in the garden wore a crusting of liquid seed pearls, and the thin-stemmed blossoms of cosmos and daisies hung down their heads with the burden of dewdrops they carried.

Thinking to lighten its watery load, I lifted a rosy cosmos blossom from behind the second-blooming delphinium spires, but I did not shake off a single drop, for in the golden center of that blossom a big bumblebee lay fast asleep.

He lay comfortably on his side in the upper cupping of the petals with his black hairlines of feet and legs carefully twined in the small jungle of stamens. His gossamer wings were spread with wet, and every black and yellow hair on his velvety body was tipped with a dew drop, minutely small, minutely gray.

With my little finger I stroked his fuzzy, dew-sprinkled back, spreading the moisture, wetting the velvet. The bumblebee did not stir. I lifted one transparent wing. There was not the slightest shadow of response. He lay as still, as insensitive, as death; but that fat little body looked too well held together by bumblebee-ness to be only a shell, and I knew he was asleep.

Besides, there were others. There were two bumblebees asleep in two yellow dahlias, one bumblebee asleep on the cushion of a golden marigold, and there were seventeen bumblebees sleeping in seventeen cosmos blooms—one in a white blossom, one in a pink, and all the rest in the deeper, rosier blooms.

The candent sun rose higher, climbing toward the tops of the trees, and the shadows in the garden retreated before the increasing light. Again I stroked the bumblebee's wet, velvety back, and this time he moved. This time he stretched his two middle legs, one out to either side, did it so sleepily I almost yawned for him since he could not, then he snuggled deeper into his yellow pillow and lay there still and quiet. A light breeze swayed the flower stalk and he slept on as in a cradle.

The sun spun higher into the sky and the first direct ray of light shone into his rosy bed. The bumblebee stretched one middle leg. Then he stretched the other. One by one, lazily, slowly, he stretched his front legs, then his back legs, until all six had been thoroughly stretched, and he rolled sluggishly forward to stand upright upon his feet.

Now he lifted his two front legs above his little elderberry of a head and slowly brushed his antennae, brushing first from bottom to top, then again from top to bottom. He rubbed his slaty eyes. All about him his fellows were awakening, brushing their antennae, vibrating their wings, drying themselves off, getting ready to go.

But the little sleepyhead only raised his wings from his back, extended them twice, raised them, lowered them, then stumbled down into the lower half of the flower, still in semi-shadow, crept into the curve of its petals—and went back to sleep!

All of his fellows were flying about, booming about the garden, but he slept on for the few minutes more until the sun grew hot upon his back and there was no more shadow in his rose-and-yellow bedroom. He rolled upright upon his feet, went through the stretching and rubbing routine once more—this time with more energy—flexed his dry, transparent wings, vibrated them rapidly, and took off into the golden morning.

These were probably all male bumblebees. They all had yellow (or white) heads—a fairly dependable indicator of their gender, depending on their species—and further, only male bumblebees have the freedom to spend their summer nights under the moon in a flowery bed.

Occasionally a worker bumblebee may be caught out by

darkness, and she will settle down for the night in the closest flower. But almost without exception, queen bumblebees, quintessential females that they are, and their more-or-less female workers spend all their nights and all their days on duty in or for the nest. All male bumblebees, on the other hand, spend their brief lives bumbling about the sunny gardens, gathering nectar only for themselves, and making love if the opportunity arises.

Male bumblebees are, truly, "only sex objects," but without them there would be no more bumblebees, and without these large bumblebees there would be no more red clover over the entire world.

This is a matter of real concern, for each colony of these large bees seems to consist of only a hundred or so adult (working) bees; and, while each colony will produce perhaps one hundred queens each summer, the number of colonies appears to be steadily dwindling. Only the young queens are able to survive the winter, and even the majority of these probably do not.

Once a young queen is wedded—not in the air like a honey-bee queen, but upon the ground or among the late summer herbage—she digs for herself a small chamber in the earth, usually not far from her home nest, on a spot that inclines toward the north, so that she will not be warmed into consciousness too early in the spring; and there she spends the winter.

When spring flowers are in full bloom, the young queen digs her way out of her hibernaculum and bumbles about among the blossoms. In these heady days of freedom she drinks her fill of nectar and drifts about in the warmth and the light—but not for long. Nature pressures her and she seeks a place to rear her own colony—perhaps in an empty chipmunk burrow, in a last-year's bird nest, or in a clump of dry grass by a fence post.

She does not bother to clean the nest, or to line it, or to change it in any way. She must prepare for her young ones, and she begins immediately to collect the fine wax which now exudes from between the hard plates that protect her abdomen. She mixes this with pollen and fashions a small waxen jar which she fills with honey as her basic food supply.

Now, with food on hand, she builds a single waxen cell. In this cell she places her first eggs, usually eight, and seals it with a waxen lid. Then she broods upon this lid, warming the eggs with her body, for the three-to-five days it takes for the eggs to develop and hatch the little pale grubs of larvae.

The queen now feeds her infants either dry pollen or a syrupy mixture of pollen and nectar. After a week of this feeding, the larvae spin their separate cocoons and settle down into their pupal state. The queen continues to brood the cocoons while she builds more cells, lays more eggs in them and gathers more nectar and pollen.

Twelve or fourteen days after becoming pupae, the new bumblebees are helped out of their cocoons by their queen mother, and, damp and soft and silvery-gray, they creep to the honey pot for a meal.

Three days later they are full-fledged, full-colored bumblebees of the worker class, full of bumblebee wisdom and ready to attend the burgeoning young about them; while the queen, who has been handling all this by herself, now settles down to laying eggs, taking an occasional flight into the open air for a sip of fresh nectar and a moment's taste of freedom.

Incredible Caterpillars

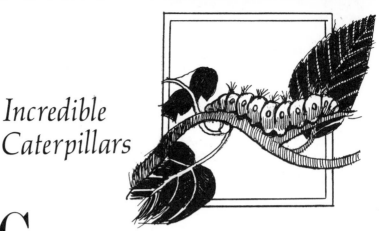

Caterpillars, those fantastic, destructive, creepy-crawly creatures that will, improbably, grow up to be butterflies and moths, are with us in superabundant forms and numbers on the full-blown days of summer. They are chewing holes in cabbages, devouring tomato leaves, consuming the ferny green of Queen Anne's Lace, gnawing curved, retreating, milk-bleeding slices from the leathery leaves of common milkweed, and gourmandizing on a hundred other species-specific foods.

Theirs is a life of gluttony, but they have only two weeks—plus or minus a few days—in which to grow from minute hatchlings to fat, full-grown caterpillars fifty times or a hundred times —or even, in some cases, a thousand times—larger than the tiny speck of creeping life that first ate an escape hatch in its imprisoning eggshell, then ate the eggshell, then began on the leaf to which the eggshell had been fastened.

To fill themselves with food is their only aim. But since caterpillar bodies are mainly digestive systems, geared to digest food just as fast as the caterpillars can eat it, they are always hungry.

Their redoubtable jaws are sharp as razors, allowing them to slice through leaf substance and tougher leaf veins as easily as they once sliced through their thin and weakened egg shells. Small mouth parts, called maxillae, hang below their formidable mandibles and help to push the sheared-off bits of food into the gaping mouth and speed it on its way. But no matter how rich the food or plentiful the supply, and no matter how fast they shove it in, they are still hungry for more, so they eat even faster.

The faster they eat the faster they grow, and, unknown to themselves, they are wearing tough, cuticle skins of only limited size. Suddenly, after two or three or four days of unlimited eating, each skin is too small for the body it is holding together.

Now, at last, the caterpillar is filled with food. It stops eating.

But after one pleasant moment of satiety, it begins to feel most uncomfortable and probably a little bit sick. It moves aside and spins a small cushion of silk to rest upon.

What with the moving about, and the bulging and pushing of its too-big body, the old skin splits over the shoulders and the caterpillar steps out of confinement—not in naked skinlessness but in a bright, new, soft and very stretchable skin which had been part of the bulky growth going on in there.

Even before it is entirely free of the old skin, the caterpillar begins to gulp air. It gulps as much air as it can possibly hold, blowing itself up like a small balloon and stretching the new skin just as far as it will stretch.

When the new skin has dried to a comfortable, too-big-for-now size, the caterpillar blows out all the air and, immediately, it is hungry again. For a starter it eats the skin it has just shed, then goes back to eating green leaves just as fast as it can cram them in.

Now it has to eat enough food to see it healthily and heartily through its present time as a caterpillar, then through its entire period of pupation, whether that be for two summer weeks or for several winter months, and—if it happens to be of a genus whose adult form does not eat at all—to nourish itself through its brief reproductive life as an adult.

Some of those adults who do eat charge their body cells with poisons from their youthful eating and are free from predation because they taste bad or cause illness to any creature who eats them. So gluttons they all are, but there is unconscious purpose in their gluttonizing.

I look at all these sausage-shaped caterpillars—the smooth-skinned ones, the hairy ones, the ones with poisoned spines—and I can see no foreshadowing of the thin-legged, airy-winged insects they are destined to become. But that is because no part of the caterpillar I now can see will become those intrinsic parts that make a butterfly a butterfly.

Hidden deep among the cells of the caterpillar's body are microscopic clusters of other cells that, by the end of the pupal stage, will have become the butterfly wings, the butterfly antennae, the butterfly legs, the butterfly proboscis. These cell clusters are alive, but they are dormant. They are called imaginal discs, not because they are imaginary but because the adult stage of any insect is known as the imago.

Even so, some of the caterpillar's other body cells probably go into the substance of these specialized butterfly organs, for even the heavy-bodied moths like the Cecropia or the Luna are smaller than the caterpillars from which they spring.

One of the strangest things about a caterpillar is that as it grows the cells of its body do not multiply. Most living things grow through the constant division and resultant multiplication of their body cells. But the individual cells in a caterpillar do not divide, they just increase in size.

This unusual method of growing may be induced by the presence or lack of some presently unknown hormone, or it may be caused by an inhibitor of some kind. And the reason for it may be the protection of those imaginal discs, which must not begin to develop inside the caterpillar body. Whatever the reason, the fat, blue-green, two-and-one-quarter-inch caterpillar of the Promethea moth has exactly the same number of cells in its body that it had when it was colored yellow-and-black and was only one-fifth of an inch long.

Some caterpillars change their colors every time they change their skin, some change it only once or twice, and some never change at all. The cabbage butterfly, for instance, keeps its green coloring through all the days of its caterpillarhood. Some of the spined caterpillars change the colors of their spines as well as of their skin, and caterpillars with rough tubercles often change the colors of those.

Some caterpillars are highly social, clustering closely together while eating, sleeping or traveling about, while others are strictly loners. But whatever their colorings and whatever their habits, it is likely that they are adaptations that have evolved through the long ages to enable them to survive in their several environments.

Different species, even closely related ones, have evolved differing ways of coping with similar problems of feeding, of weather and of predators.

The caterpillars of the common cabbage butterfly are green, they live apart from one another, and they burrow their separate ways into hiding in the head of cabbage they eat.

The caterpillars of the large white cabbage butterfly also feed on cabbage leaves, but they feed, quite visibly, on the outer leaves. They keep together in a cluster and wear varied colors throughout their caterpillar lives. After their last molt they are yellow with purple stripes.

Birds feed eagerly on the green caterpillars of the common cabbage butterfly—when they can find them—but the caterpillars of the large white butterfly are left strictly alone, so they can be just as bright-colored and as reckless in their behavior as nature permits them to be.

The monarch butterfly and the milkweed moth are not close relatives, but both species feed on the thick green leaves of the milkweed plants. We have learned that a poisonous substance from the milkweeds is present in both the caterpillars and the adults of the monarch butterfly, and that this poison is their protection against being eaten by the predacious birds all about them.

Presumably, then, this poisonous substance is also present in the caterpillars of the milkweed moths who feed just as voraciously on the same plants. Yet the colorful, smooth-bodied caterpillars of the monarch stretch themselves out, all alone, and feed on the exposed upper side of the milkweed leaf, while the extremely hairy, dull, burnt-orange-and-black milkweed moth caterpillars feed in pairs on the hidden underside of the milkweed leaf.

Probably the greater number of caterpillars who must hide from birds solve their problem by coloring themselves some shade of leaf-green or of twig-brown, but the hawkmoth caterpillar has large, brightly-colored eye-spots that are supposed to frighten off any attacker, especially when it lashes its body from side to side in snake-like motions.

Some caterpillars have poisoned spines, and others, the swallowtails, for instance, have retractable, threadlike horns that give off odors offensive enough to repel any creature with ideas of harming the otherwise helpless caterpillar. Others, like the orange dog and the hickory horn-devil may, with their varied colorings, be using the broken-pattern method of camouflaging themselves from the searching eyes of hungry birds.

But these tactics and strategies, these color changes and these habits of behavior are all built into the juvenile caterpillar. The caterpillar is not consciously aware of them—neither of the things that just happen or of the things it does. In its dim consciousness the caterpillar rarely senses even so large a creature as a human being who stands close beside and lays a hand on the leaf it is eating. The caterpillar is scarcely aware of the leaf, much less of the plant it inhabits.

I ponder. And I ponder deeply. What magical secret ingredient is hidden in those imaginal discs so that the butterfly emerging from the caterpillar skin should be infinitely more alert, alive, and aware than the caterpillar it so recently was?

Bob-White Tactics and Strategies

With my new puppy, Kuonny, I followed the foxes' path through a fallow field filled with goldenrod, milkweed, Queen Anne's lace, and all the colorful ragtag and bobtail vegetation that springs up in corners of untilled land. A family of Bob-whites, two parent birds and eight or ten little ones, burst into the air in front of Kuonny's nose. He leaped after them, but my hand tightened on the leash and he stood by my side.

The mother bird dropped out of the covey and began a pitiful broken-wing exhibition about twelve feet ahead of us, moving away from her family, drawing our eyes to her. When we made no attempt to follow her but only stood watching the show, she abandoned that ploy and began to fly in short, close circles around us. She flew well within leash-length and so very slowly that I couldn't understand how she stayed airborne.

We still did not move. She then began a highly excited back-and-forth, flying, falling and calling maneuver off to our right. I dislike to upset any creature, so I struck off into the field away from the excited mother. But she changed direction too, and swung out ahead of us, fluttering helplessly just beyond our reach until we had been, from her point of view, successfully diverted and she could safely slip away to find her well-secreted family.

The wiles of parent Bob-whites in their efforts to lure possible predators away from their young seem to me far more varied and more keyed to the circumstances than the humbuggery of other birds, but authorities say that although injury-feigning sometimes looks as though it is a case of reasoning, it is too common a habit among too many species of birds for that to be so.

I certainly do not consider that birds think their ways through their short lives, nor that they conceptualize as humans do; but, even if instinct (a word I distrust) orders their every move, it seems to me they have, now and again, to exercise judgment. If a bird makes a mistake and does the wrong thing, surely that indicates that it had the power to make the choice in the first place. . . .

Several years ago on a September day I watched a distressed male Bob-white struggle through the hay stubble with his rear dragging so hopelessly on the ground that I was certain he had received a fatal load of buckshot from dove hunters in a nearby cornfield. He could scarcely make it to the honeysuckle of the fencerow, but, once there, he was miraculously healed. It was an entirely successful ruse. He had fooled me completely. But then he made a blunder.

The very instant he was safely concealed in the tangled vines he changed his weak cries of distress to a low calling, and, from directly at Kela's and my feet, fifteen tiny, down-covered, one-day-out-of-the-egg Bob-white chicks rose and ran, waggling and bobbling like buffy, earth-bound bumblebees through the impeding stubble, to reach the care and protection of good old papa who was, I hope, trembling in horror at what he had done.

Not that there was any shortage of Bob-whites—then. During those years the cheerfully confident "Bob-white!" whistles of spring and summer, and the wistfully sweet covey calls of fall and winter sounded almost constantly from hills and fields and woods and meadows all about us.

And their nests were everywhere: Along fencerows, under stumps at the edges of woods, under bushes in the pastures, and in heavy clumps of grasses or weeds. Many were hidden along the roadsides where highway department mowing machines often exposed the ten-to-eighteen pure white eggs and sometimes removed the brooding bird who remained too steadfast to her duty. Instinct probably ordered the hiding of all these nests, but surely individual decisions placed them.

Bob-whites are easy birds to recognize. With their heavy, rounded bodies, short wings, stout legs and bills, they look like small escapees from a chicken yard. Their calls unequivocally announce their identity. Furthermore, the sexes are clearly differentiated and you know which bird of the pair you are watching.

The male has a white face and throat, and his bill is almost black. The female has a buffy face and throat, and her bill is a light horn color. Her feather colors are less intense than his and she has less black on her breast, but that requires a side-by-side comparison, while the face color is distinctive all on its own.

In those past years of great Bob-white populations, several coveys, of fifteen or twenty birds each, came regularly to the feeding station in my backyard. Many other coveys had regular territories in various sections of the countryside that Tiki or Kela and I so regularly patrolled, and we came to know where we were likely to find certain Bob-white families.

Often we discovered the small circles of their droppings, which betrayed where they had spent the night sitting comfortably upon the earth in a warm close circle with heads facing out to the round horizon. Once in a while, if we were very early upon our rounds, we could catch them still abed and thrill to see, and hear, a covey explode to the four winds as each bird took off in the direction in which it was facing.

Once, in those long ago days, I witnessed an incident that almost had to be both organized and planned—but I'm sure it was neither. On that day I watched, from my dooryard, as an unprecedented host of Bob-whites quietly mobbed a cat.

It was my own cat, my sweet-tempered black one, when he was scarcely more than a year old. He had been walking along the crest of a rise in the woods playfully scattering autumn leaves about him, when suddenly, six or eight feet in front of him, Bob-whites began to rise—four, eight, ten at a time—from the loose carpet of leaves on the woods floor, and to walk, slowly and steadily, in close formation, directly toward the cat. The only sound in the whole woods was the steady patter of their dozens of pairs of tiny feet in the loose and crackly leaves.

The cat stopped in his tracks and stared into the faces of that unbelievable number of advancing birds. His hair began to rise. The silent birds walked straight toward him. The cat's ears went back. More birds kept rising from the leaves and joining in the push. The lead birds were not a foot away from him when the cat turned his back upon them and tried to walk off unconcernedly in the opposite direction.

But the birds kept determinedly following. They walked steadily, silently—except for the patter—just six inches behind his little black heels, in a bobbing brown phalanx of sixty to seventy birds. The cat laid his ears back, lashed his tail, and his white teeth chattered as he walked back along the crest of the ridge.

The birds followed. Too closely. Too relentlessly.

The cat turned west at the woods road and came down the ravine. The army of rounded brown backs followed at his heels. He dodged around stumps. He jumped over logs. Steadily, inexorably, the unhurried birds came on. The cat's ears fell forward, his eyes shifted from side to side, his tail swung low.

With a cry of anger well mixed with fear, the cat dashed out from the trees, up across the lawn, into the house and to his fortress behind the sofa. And the legion of mute, unruffled birds vanished as they had risen, into the dry scattering of autumn leaves on the woods floor.

The Song of the Hairytail Mole

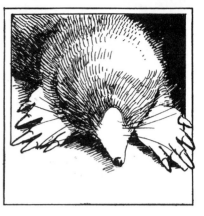

Kuonny killed a mole two days ago, a hairytail mole, out on the hillside above the pond where the earth is sandy-loamy and the timothy grass grows thick-stemmed and only medium high.

I could have stopped him, and I should have, but Kuonny is still a puppy and not adept at catching mice and moles among the stems of grasses. He pounces upon them with close-held feet. He pins them down. But they flatten their bodies and flow away beneath the thick green stems that somehow always lie between their furry backs and Kuonny's straining nails.

And so it was with this one. It slipped away easily, but again and again, too many times, until, at the end, I stood so immersed in its death-song, so detached from its death, that I watched without awareness the white puppy-teeth nip round its throat, the dead body dropped carelessly to earth.

The mole was a little one, the smallest of our true moles. Its slatey fur was rich and shining, and its short, round tail was as smoothly sheathed in fur as its body.

Its dainty hands were turned palm-out, the palms soft and rosy-pink as a baby's hands, so square, so fine, so weak and useless-looking in death; so clean, the claws so white, as though they'd never dug twisting networks of earthen tunnels under the living roots of the grass.

The mole's nose was bare and rosy-pink, wedge-shaped and used mostly as a wedge, for it could only smell things that lay within two or three inches of its tip. But an essential organ for its breathing, nonetheless.

Small depressions, like narrow slits, on either side of its face marked the places where its almost useless, embryonic eyes lay hidden under its fur. The living mole knows light, and darkness, but nothing more through its actual sense of sight.

I could not find its ears, for there are none; only a channel opening, like a pore, at the surface of its skin—and this channel can be closed at will. Perhaps it was closed. Perhaps I did not know how to look into that thick velvety coat. Perhaps I did not know what I was looking for.

Without the cupping of the outer ears, some authorities say, the hairytail mole cannot hear at all; but others say its hearing is acute, that it can hear the faint commotion of a beetle falling into some distant portion of its hunting tunnel. Ah, yes, says the disputing voice, but it does not hear—it feels. It feels the vibrations of the beetle's fall and the vibrations of the beetle's attempts to right itself. It feels these vibrations in the air and in the soil. It feels them with its entire body, but especially with the tactile organs of its nose, its hands, its feet, its tail.

But whether the hairytail mole hears these vibrations or feels them, it travels directly to the beetle—and eats it.

For the hairytail mole is always hungry. It is active day and night. It may eat three times its own weight in worms and insects in every twenty-four-hour day and will quickly die if its food supply dwindles. For all that, it seems not to live so briefly, so frantically or so antisocially as its fellow insectivore, the shrew. The apparently calm hairytail lives for four or five busy years and quietly shares its meandering tunnels with several successive generations of its own family.

This extended togetherness does not result in the nightmare neuroses of ballooning meadow mouse and lemming populations, for moles are not very prolific. The hairytail mole does not mature until it is ten months old, and it only produces four young at a time, although it may, possibly, have two litters a year.

This could make for a fair number of moles, it's true, but moles, like mice, are low on the food chain and are eaten by foxes and cats and dogs and hawks and owls and skunks and crows and whatever else has a yen for small prey, so that the population of hairytail moles is rarely as many as ten or eleven per acre of dry pastureland, where they aerate the soil and eat grubs and eggs and mature insects in vast and valuable quantities.

It is the inborn duty of every hairytail to dig; so their tunnels, lived in for many years by successive numbers of moles, become intertwining and complicated, with galleries, rooms, hunting tunnels, and escape hatches in a bewildering array. But the system of tunnels belonging to any one family is confined to approximately one-fifth of an acre, and it rarely reaches further than eighteen inches below the surface.

The tunnels at or near the surface, where the loosened earth is pushed up and out to form molehills, are temporary tunnels. Permanent tunnels are deeper and are made by compacting the earth, by pushing and firming it all about the busy, digging body, and there is no earth to be thrown out.

To do this, the mole pushes its wedge-shaped nose and head into the earth before it, while its hands, whose palms are turned outward, scrape and push and pull upon the earth around it. It is as though the mole were swimming, using a breast stroke, through the dark deeps of the earth. When it is digging a temporary tunnel, its hind legs keep kicking out the loosened earth behind. When the tunnel is to be permanent, the hind legs are crouched beneath the body, helping to compress the earth below.

The hairytail mole spends a great deal of time on top of the ground, too, hunting among the weeds and the grasses, creeping about the bases of their stems, exposed, in its blindness, to all the dangers of the open air—and it was here that Kuonny caught one.

There was the puppy's initial pounce, high and rounded, then a series of quick bounds and side-steppings, and suddenly the mole began to sing—and I was startled into immobility.

The mole's song, rising from the bases of the grass stems, was sweet and pretty. A high, rich treble, it was like a canary's song, or a sweet-voiced sparrow, with every note separate and distinct. I stood there on the hillside, my eyes wide open, the morning sunlight streaming about me, and I heard that song, scarce believing. For the space of one breath I stood in moonlight at the foot of mountains and listened as a dying Indian, in dignity and calm, threw away his song.

The mole's song ended. In the silence, Kuonny's teeth closed around the little throat, and he pulled the lifeless mole out into the sunlight and dropped it there.

This afternoon Kuonny cornered another hairytail mole under the peonies in the backyard. By degrees, he worked it out into the grasses, losing it, finding it, losing it again. The black cat joined him in the unfair game of hide-and-seek. Two against one, now, and the black cat quick and experienced.

Suddenly this mole, too, began to sing. But this song was a high warbling trill with the notes all running into one another. This song was neither rich nor calm. This song had anger in it, and defiance, and it trilled on and on as the cat and the dog pursued the mole in zigzagging courses through the grass.

There was a sudden rush and a skirmish, a minute-long scramble beneath a lilac bush. The cat came out, all pretended nonchalance, but the dog was under the bush frantically digging. That hairytail mole got away!

Until this week I did not know that moles could sing. Within three days I heard them twice, and unmistakably. The treble warblings of the two moles, under almost identical circumstances, were strikingly different—and so were the ends of their stories.

The
Pearl-Makers

On a sunny November afternoon I worked at pulling clumps of red dead nettle from a section of the vegetable garden. The hay mulch had disappeared from this area in midsummer and I had not renewed it, so I was reaping the rewards of my negligence.

In place of summer's debilitating heat, though, I worked in crisp, bright autumn weather. No gnats, mosquitoes, flies or yellowjackets buzzed around me, or crept about on my skin, or irritated my flesh. The garden earth had reached that state of perfect friability it attains in late fall, so that I rejoiced to work with it; and the dead nettle, allowed to grow unchecked, had crowded itself out so that instead of hundreds upon hundreds of small plants to be dealt with, I had only dozens upon dozens of healthy clumps to root out.

Moreover, this was to be a new learning experience for me, although I didn't know it yet. I was just pleasantly occupied in pulling great, wide weeds which came up without my expending too much effort and which, when pulled, left so satisfactory a clear area behind.

Now in the season just past, my garden had become infested with slugs. They had hung on the tender lettuce and new green peas each evening in unbelievable hordes and had rasped away leaves and blossoms with such single-minded devotion and appetite that there was nothing much left for the gardener to enjoy.

My efforts to dislodge them had been futile. They apparently devoured dustings of wood ashes and powderings of lime, they crept undeterred over scattered sand and broken egg shells, and they wouldn't be caught dead in a saucer of beer.

They were not devouring my garden out of malice. They were simply following the rule of living where the eating was easiest. I had, in a sense, invited them in by providing an unnatural concentration of unnaturally tender plants in a small and unnaturally clear area. Besides, the hay mulch (over newspaper) gave them the coziest of living quarters.

But I hadn't, this fall, seen a slug for several weeks nor given the species any thought, and so I was quite surprised when I pulled up the roots of the red dead nettles and found small cavities in the earth around them in which great numbers of slugs were hiding away from the chill of autumn weather.

Slugs would need to get into snug quarters, for they are so bare and so wet they must be extremely susceptible to cold; but how those little, soft-bodied things could form those cavities, I do not know, unless they did it by pressure of their bodies on the light, loose earth.

Scores of the cavities I found had no slugs at all in residence but were filled with small grayish or pinkish pearls that came rolling over the ground as I pulled up the plant roots, and spilled them out like jewels from their caves. They were truly exquisite —hard, almost translucent, alive as a pearl looks alive but even more truly so, and just-not-quite-perfectly round.

They were nearly one-sixteenth of an inch in diameter, and there were from twenty to thirty of them in each little hidden cave. They had to be slugs eggs, I thought, but, while they were moist, they were not at all mucus-y. Each one was free from the rest and they rolled easily and singly about when I upset them from their cache.

Any faint doubt I may have had about the origins of these eggs was dispelled when I came upon a slug in one of the cavities, shedding small and lovely pearls from beneath her mantle. By twos and threes and fours and fives the eggs simply appeared at the line where body and mantle join and were slipped off onto the earth by apparent muscle contractions. I had the feeling that these eggs had been carried under the mantle for some time before being shed, but I don't know that that is the case.

I say the slug was shedding eggs from beneath "her" mantle because it seems to me that at the instant of shedding eggs the animal is female. Slugs are, though, hermaphroditic, and are, I understand, one of the few animals who are capable of complete self-containment. However, they apparently do not avail themselves of the opportunity, but hang about lawns and gardens in their twosomes, keeping the world supplied with slugs without resorting to innovations.

As a naturalist I am forever remonstrating, "But snakes are not slimy," "But frogs are not slimy." However, I will say, heartily, "Garden slugs are surely the slimiest animals on this green earth." Like snails, they exude a mucus beneath them on which they can travel smoothly and easily over all sorts of terrain. But garden slugs also exude mucus over their entire rubbery white, yellow, pink, lavender, purple, brown or black bodies, and they *are* slimy and unpleasant to handle.

Not all of the slugs, on this sunny November afternoon, were hidden underground. Some were out on the earth surface under the protective covering of the red dead nettle clumps, and others were up on the stems and among the leaves and branches so that sometimes, when I pulled a plant, I also crushed a slug or two in my hands.

The first time this happened I wiped the soft body from my fingers, and as I did, I felt a hard, flat substance there. Surprised, I washed it clean and found that I held a small disc, pale colored, but delicately touched with rainbow tints like the surface film on a roadside puddle, like mother-of-pearl.

And mother-of-pearl is what it was. I had forgotten, though I must have known, that a slug is simply a snail that has learned to live without the protection of a spiral shell. Their bodies, though, have not yet quite learned how to stop, completely, the forming of the shell substance, and so each one carries within it a tiny disc, the final vestige of the shell that once housed its ancestral kind.

But what *good* are slugs? someone demands. What do they do besides destroy vegetation with their rasping sandpaper tongues?

They are only eating. But sometimes, as with the human population, there are too many of them. I do not hold with those who think that every thing on earth is here to serve mankind— although, in the long run, they may do so, even in the concrete terms usually meant.

Slugs are little grazers who eat green plants which alone on this earth translate the sun's energy into food and growth and life. Slugs are eaten by snakes and toads and turtles, and, possibly, by birds. Some of these animals, in turn, are eaten by owls and skunks and dogs and foxes and the families of cats, which return their bodies to earth. And so it goes.

Actually, life is good for life—the richness of its texture, the variety of its forms, its intricate fitting into niches, into niches, into niches. . . .

Foxes on the Hilltop

The morning was gray and chill with a light, intermittent rain falling on the leafless trees and the browning winter grasses. The thirty or so holsteins which usually, at our walking hour, are grazing on the hillsides and along the main stream of the pasture, were off today at the far end of the marsh, meandering slowly toward the barn, so that Kuonny and I had the whole magnificent sweep of this broad pastureland to ourselves.

Kuonny, at the end of his twelve-foot leash, seems always to feel free and curiously unhampered. But I, because of the leash, am responsible for his safety, and I must keep a sharp lookout for the closing-in upon us of determined mobs of dog-hating cattle. For whatever noble characteristics these animals have lost in the demeaning process of becoming domesticated, they have retained every atavistic facet of the protective technique called "mobbing the wolf." They ignore me and concentrate on the dog; and the situation sometimes gets sticky.

But today the entire herd was headed for the barn, and I slid beneath the barbed-wire fence and followed Kuonny through the soupy marsh, as light-hearted and as unconcerned as he, giving my whole attention to watching the tussocks for cottontails, meadow mice or muskrats, and to the water borders for woodcocks, herons or a possible mallard.

We waded the swollen, swift-flowing creek where it spreads wide and—usually—shallow beneath the oaks, crossed the narrow strip of marsh on its far margin, and side by side, began the climb up the first rise of this long sweep of hills.

A flicker of movement on the western crest of the hills caught my eye, and I froze. Kuonny, running with his nose to the ground, but well-schooled by now, froze beside me. He lifted his brown face and studied mine with puzzled eyes.

For a moment I considered whether the two creatures up there under the sassafras trees were two dogs at play, but the sinuous, free-flowing grace of their body movements and their long, full-furred, trailing tails labelled them as foxes.

I sank to my knees and Kuonny to his haunches. This movement caught the foxes' eyes. They stopped their wild tussling and crouched in the grasses to stare down the hill at us. Why they did not see us sooner, I do not understand. We were fully exposed to them as we crossed the creek and the margin of the marsh, and were not much hidden from them for ten minutes before that. Unless, of course, they were then climbing the hills themselves. Their backs would have been turned toward us. . . .

At any rate, they stared down at us now, only slightly suspicious and not very much alarmed. I was dressed in an earth-stained gray-blue rain suit which covered me from crown to ankle, and Kuonny was wearing his best brown and white. If foxes see only in shades of gray, as dogs supposedly do, our lumpy shapelessness, from that distance, must have looked to them reassuringly like a reclining cow. Even Kuonny's occasional head movements were apparently accepted as a normal part of the scene, and the foxes went back to their wildly careening play.

One fox, though, played hesitantly. He was not quite sure, yet, of our identity—something about that reclining cow was not quite right—and he kept stopping in mid-action to stare down the hill at us. But the other fox nibbled at his ears, raced around him in tight circles, licked his nose, enticing, teasing, cajoling; and soon, since we made no threatening movements, both foxes were wholeheartedly dodging and weaving and chasing each other up on the top of the long pasture hill.

This was play at its exuberant best. Racing, then rising on hind legs, balancing with front paws on each other's shoulders, swaying together, mouth grappling with mouth for the better hold. Down on all fours again, circling and weaving, leapfrogging one another's back, rolling together in the grass. Then up on their hindlegs and grappling again. Over and over the swift muscular maneuvers. Over and over the free animal fun.

By this time Kuonny had discovered them and he watched as absorbed and as silently as I. Minutes went by. Five minutes. Ten minutes. Fifteen. Still the foxes tussled, raced, grappled, and rolled. Kuonny stretched out comfortably on the ground, chin on his paws, watching the play; and I settled back, sitting on my heels, delighting in the total giving, the complete abandon of the foxes on the hilltop—amazed at the quiet serenity of my watching dog.

These were red foxes. Healthy, well-fed, and surely not much more than one year old. Their coats looked sleek and full, faintly reddish-yellow but much, much mixed with brown, and the tips of their tails were more yellow-brown—clay-stained perhaps— than the classic white.

Their play grew wilder, faster, and suddenly, with glorious banners trailing, they were racing, side by side, obliquely down the long sweep of the hill, directly on a line with us. Faster they came, and faster, shoulder to shoulder, over the rising, falling ground.

Every fiber of my body exulted, shared, yearned, swept with them down that rushing wind. I felt the wave of excitement pass through Kuonny's body, and fearful that he might dash upon them, I said "Stay!" under my breath. Not the hissing sibilant, not the forceful consonant, simply the vowel sound, "ay," soft in my throat.

But the foxes, a thousand feet away, joyously racing and off their guard, heard the fearful sound of the human voice, and they stopped on the instant.

Shoulders together, they stared for one moment at the peculiar creature on the ground below them. As one body they whirled, and together they fled straight away across the field toward the woods, toward the precarious safety of their hidden den.

As they ran, now, one was just slightly to the rear and to the side of the other. They ran just as fast, I suppose, perhaps faster, than in their headlong race down the hill; but this time they ran in fear and not in joy—and that made all the difference.

Picnics at the Roadside

The country neighborhood in which I live has become so effete that we now have a garbage-trash collection twice in every week. However, the great mechanical monster that devours our leftovers rolls by between three and four o'clock in the morning so that any offerings for its capacious maw must be set out at the roadside on the evening preceding each designated collection day.

Since the end of my lane is just beyond the edge of my woods, wandering raccoons were quickly attracted to the temptingly odoriferous cans of refuse I left out there, unattended, the whole night long.

At first, and this was during the hunger-month of February, the raccoons mostly came to the cans one at a time. Last year's youngsters were out on their own and their groupings were so divided by death and disaster that each survivor was left to hunt by himself. Also, the mating season was practically over, males and females were losing interest in each other, and they now traveled, mainly, in self-sufficient singleness. And, if there were early babies in any females' nests, they were left behind while their mothers set out alone on their nightly search for sustenance.

During this quiet time of year a lone raccoon made little commotion around the garbage cans at the roadside. Usually the only sound I heard was a metallic ringing when the lid of the can dropped onto the frozen ground or hit the edge of the blacktop road. A little later there might be a thud and a clatter when the can tipped over, spilling its waste contents, and the foraging raccoon before it rolled to a stop.

Sometimes a second raccoon came by for a share in the booty while the first finder was still there. It might be greeted with a low growl, or with a snarl, or with no sound of acknowledgment at all. Both raccoons would pick about among the offerings, amicably ignoring one another—until both wanted the same thing at the same time. Then a brief fireworks of snarls and screams would erupt from the roadside, but the dominant raccoon would soon make off with the prize, and the other would resume his quiet scratching about in the trash.

But by mid-spring, with acres of forest providing dark cover for their comings and goings—and offering a safe refuge to dash into if danger threatened them at the site—mother raccoons began to bring their two or four or six babies on some of their very earliest feeding trips away from the home den. Then, as spring turned into summer and the babies grew older and bolder, the road edge became the scene of noisy and exciting picnicking with many a fierce battle waged over the sparse and delectable tidbits to be found in that regularly reappearing cylinder.

As autumn approached, one mother raccoon with four roly-poly youngsters tumbling about her feet became the regular "first feeders" of each garbage-can evening. They came shuffling out of the woods as soon as true darkness had fallen and they would be indistinguishable except as motion among the darker shadows, or else clearly and delightfully visible in the goldust-light of the late summer moon.

As she approached the garbage can the mother raccoon always gave a loud grunt which I translated as satisfaction that the can was in its rightful place. She followed that with a churring sound which collected her family about her, as, with agile fingers and amazing strength, she opened the most ingenious of the fasteners I was able to devise. Her churring was loud, somewhat like the "talking" of a great cat, and, once the can was up-tilted and its contents spilled and spread, her little ones joined in the churring, adding occasional individual trillings in their excitement.

But these happy churring sounds and the very occasional feeding silences were interrupted, as often as not, by hissings and snarlings and sharp, wild, staccato cries as the youngsters quarreled over a chicken bone or the scanty leavings in a flattened catfood can.

If one of the little ones was hurt in the melee it cried piteously in tones that hurt my heart and that brought the mother raccoon to scatter the battlers, succor the injured and, perhaps, to take over the prize for herself.

But, altercations or no, the churring was repeated almost con-

stantly during the feeding, and, while it probably indicated comfort and contentment-while-feeding, it seemed also to be something like a "covey call" that kept the young ones from absentmindedly wandering off into the wilderness by themselves.

If a lone male raccoon came by wanting a share of the spoils,
he was usually driven off by the combined hisses, snarls and
growlings of the entrenched family. If another raccoon family
approached, both families first went through stylized and hostile
maneuvers and then settled down to feed and spat together.

However, since all vegetable refuse went into my compost
pile, and I do keep a fairly careful kitchen, there was never a great
amount of food out there for the raccoons to dally over. The first
feeders usually cleared the can and the next raccoons to arrive
were left to sniff the odors of what they had missed.

Often, when the "first family" had cleared the trash of everything edible and were making their way in a shuffling group back
through the woods toward the creek or down through the over-
grown fields toward the marshes they "sang." They sang, unques-
tionably, a "song." Their tones were strange and sweet, had an
up-and-down, quavering quality, and were interspersed, irregularly, with clear, quick, star-shaped notes from one or several
voices at a time. It was truly a gypsy song for it tempted me, often,
into following them off into the night.

I truly enjoyed this opportunity to watch the antics of these
raccoons and to learn to know some of their calls and their songs;
but, after a year or so, I also grew immeasurably tired of clearing
up the scattered mess they left behind them at the roadside. So
when my visiting brother sensibly suggested that I place the
kitchen scraps in some other spot and leave the trash can free of
edibles, I hastened to take his advice.

I began, each evening, to place the small bits of table scraps
at the edge of the woods behind my house, where the animals who
came were visible from my kitchen window. But so many and so
various were the wildlings who responded that I soon graduated
from mere food scraps to small handouts of dried bread and dry
dog food, and my compost pile is notably lacking, these days, in
any fruit and vegetable remains I think my hungering guests might
favor.

But I try never to offer so much at one time that I put any wild animal in danger of becoming my dependent. It is still pretty much a matter of tidbits and only the first to come are served. Fortunately for me, as observer, many species have been the first to come.

There have been raccoons, as before, of all ages and all sizes and all degrees of experience in the great wide world. There have been skunks, and opossums, and foxes, and chipmunks, and squirrels. There have been thin little cats without a home to their names, and well-fed dogs, from good substantial residences, out scavenging the countryside.

There have been skunks bearing all the proper markings on their long and silky fur, skunks with only a tiny white cap and a smidgeon of white on the tip of their tails, skunks with one wide white stripe down the middle of the back, and other skunks wearing various combinations of all these styles.

There have been opossums fat and sassy, and opossums that were puny and shy and hungry and thin; opossums well-furred, long-tailed, and tissue-paper-eared, and opossums with tail tips lost or half the tail missing, with one ear frozen, or both ears crumpled, or both ears gone.

There have been foxes thin and ill and scroungy; and foxes fine, fluffy-furred, wild and strong; foxes terrified and snatching food on the run, and foxes eating as serenely and as calmly as my dogs.

Once, on a snowy winter's night, I watched a gray fox and an opossum contentedly crunching dog food while standing side by side against the fence. When the food was gone, the gray fox trotted off through the woods in one direction and the opossum wandered off absentmindedly in another. And I beamed upon the unexpected demonstration of woodland coexistence. Only a hungry fox kills an opossum.

The Waters of Heron Pond

The year moves toward the spring and, on this still-winter day, the edges of my neighbor's pond, Heron Pond, are free from ice. The northering sun warms the shallows in the pond's upper reaches and in the two small coves at either end of the earth dam, but in the center of the pond, above its coldest depths, a great floe of ice, some five inches thick, lies awash in its restless waters and tones down the effects of the beaming sun.

Most of the time the force of the current flowing through the pond holds the forward edge of the ice floe against the side of the iron grating that guards the opening of the overflow pipe. But sometimes, as now, the force of the wind is greater than the force of the current and the ice is pushed away from the grating and into the warmth of the upper shallows. But the wind does not hold steady and the ice raft sloshes forward and back as it responds first to one propulsion and then to the other.

All this motion in the water is erosive to the ice, and the flow of the current over the ice as it pushes against the grating wears it away even as the dark iron melts its forward margins into delicate and fanciful edgings. This battle of wind and ice and current is interesting to watch, but the slow dissolving of the ice-body, as well as its parasol effect, must keep the surface waters of the pond cold and light, and so interfere with the mixing of the water layers so essential to the life and health of the pond.

For the pond itself is an organism and upon its life and health depend the life and health of the myriad of organisms which live within its depths, its shallows, and on its margins. And this pond is ailing.

Two months ago, twenty-four hours of down-pouring rain filled the inlet creek to its bank-tops. The rush of the flood waters swept tons of fine brown silt from the creek bottom and more tons of soft earth from its banks into the blockage of the pond where the overflow pipe and the spillway together could not carry off the swirling, mud-laden waters. And the pond, too, filled beyond its banks.

The water, then, receded quickly as the air temperature plummeted, and in a few hours the pond banks and the sand bar—now a mud peninsula filling most of the shallows at the upper end— were above water level but covered with several inches of swiftly freezing mud, while the ice, forming almost as fast on the pond surface as on its banks, was at least as much mud as it was ice over a wide margin all the way around the pond.

The water level in the pond continued to drop, of course, for the next few days, dropping perhaps another three or four inches, and the center ice sagged as it thickened. But the elastic ice on the margins, strengthened by its content of earth, did not break away. It pulled apart now and then, but the fissures quickly sealed, and the margin ice developed a decided downward curve that held through the winter while temperatures rose and fell and the pond remained solidly covered with ice.

But under the warmth of the returning sun, the dark, earth-laced ice of the margins, absorbing the heat, quickly melted away. This left the sagging ice in the center free to slosh about the pond, half-covered with the water it held in its faintly concave surface, shielding ice-cake and pond alike from the desperately needed warmth of the sun.

For water, in a pond or elsewhere, is a strange and wonderful substance that, unsurprisingly, becomes heavier as its temperature is lowered, but only until it reaches a certain point; and then, miraculously, it reverses the trend and grows steadily lighter. And that certain point, most fortunately for mankind and for all life on earth, is thirty-nine degrees Fahrenheit, four degrees Centigrade).

At thirty-nine degrees water is, under natural conditions, at its greatest density. When its temperature drops below that point it begins to expand, and when it reaches freezing temperature, thirty-two degrees Fahrenheit, it expands rapidly.

Surprisingly, and in spite of appearances, unfrozen water at thirty-two degrees is much heavier than the solid ice beside it. This has to do with the crystal structure of ice and the arrangement of its oxygen and hydrogen atoms into bulky molecules, but I'm not going into that, now—or, probably, ever.

This factor, of ice being lighter than water at 32°, is only one of the great life-preserving qualities of water, but it is the quality that prevents our lakes and our rivers and most of our ponds from becoming solid, sterile, never-thawing blocks of ice.

In the fall of the year, as the air temperature drops, the water temperature drops also, although more slowly, and the water on the surface of the pond, in contact with the air, cools the fastest. As it cools toward thirty-nine degrees it grows heavier and drops through the lower layers of warmer water so that, by the time winter arrives, there has been a pretty good mixing-up of the pond waters and the *fall turnover* has been accomplished. Then the surface water, cooling rapidly, becomes lighter and lighter until some of it freezes into ice, and the top layer stays on top.

In the spring the same thing happens, only somewhat reversed. As the ice melts and the temperature of the surface water rises, approaching thirty-nine degrees, its increasing density causes it to sink through the colder, lighter waters below it and, thus, to bring about the *spring turnover* of the pond waters.

Both in winter and in summer the lower depths of a pond tend to stagnate, to become depleted of their oxygen, and, twice a year, the interplay of densities and temperatures turns the water upside down and keeps the water body alive. Even in a pond such as this one, with the current of a creek streaming through it, this turnover is vital. For the current through any sizeable body of water, including a river, does not often stir up the bottom layers of water. The actual current in pond, lake or river flows under the surface and above the middle depth, while the lower levels are rarely disturbed and simply muddle along.

The winds of March and the gales of November whip extra quantities of oxygen into the waters they lash. They beat the waters, stir them up, cause disturbances to greater and greater depths, and exert their own importance in the spring and fall turnovers.

All winds that disturb the waters add oxygen to the pond and probably mix its upper layers to a certain extent. But, in the spring and in the fall, the rising and the falling of the water temperatures, the increasing and the decreasing of the densities, the general instability of the whole water body, all these forces and conditions increase the effectiveness of stormy winds in bringing about a wholesale intermixing of stratified layers and stagnating molecules. Some biologists maintain that without the action of the winds in these crucial weeks a complete turnover would not happen and the pond would slowly die from lack of oxygen in its depths.

So I worry about the effects of that ice floe out there in the middle of the pond, about its interference with the wind and its interference with normal temperature changes. And I worry about the hibernating turtles and frogs and the myriads of smaller creatures now buried far beyond their normal depths under the mud carried in on the winter flood.

I worry. But I also know that the waters will warm, that frogs will call and turtles will rise and whirligigs will spin, that dragonflies will zoom, and the voices of the redwings and the meadow larks will announce to me, and to all of the world that, once again, and in spite of all, spring has come.

The Woodcock's Springtime Gyre

Early in my first April at my present home, I sat on a stone at the western end of the house and watched the sun flatten itself on the horizon, then swiftly slide away into the haze of a rose and lilac evening.

Bird voices at the woods-edge and in the orchard faded to chirps and to faintly heard vespers while massed choirs of frogs rejoiced in the marshes. A peculiar, short, rasping "beezzp" like that of some giant, wing-filing insect or like the sputtering buzz of crossed electric wires—though a little more stridulent or more nasal than that—kept repeating from constantly changing positions in the weeds of the field at the edge of the lawn.

I don't know for how long the "beezzping" went on before a chunky brown bird shot from the grass not twenty feet from me, cleared the five-foot fence with an inch to spare, and sped off on a great circle accompanied by an amazing twittering sound, his wings vibrating at a speed, it seemed to me, just less than a hummingbird's.

The twittering sound continued as the bird spiraled up and up, each speeding circle narrower than the last, until his dark body vanished into the duskiness of the sky. A moment of silence and then the most glittering starburst of bird notes I have ever heard came raining down through the darkness, lingering like tinkling crystal over the housetop and the meadows.

Another moment of silence, and then the bird dropped into sight against the rosy light of the sky, plunging to earth in three spectacular arcs that brought him down at almost precisely the spot from which he'd taken off.

With scarcely a pause for equilibrium, he puffed out his chest, lifted his short, blunt, white-bottomed tail, pressed his long bill down against his breast, and, with his wingtips dragging on the ground, he strutted off into the weedy field; and I heard him

uttering those nasal beezzps at what I assumed were regular check-points on his circular footpath.

On the most distant circumference of this dancing circle his voice was all but lost to me in the jubilant voices of the frogs, but on its nearest curve the beezzp was loud and clear with the bird strutting dandily right across the lawn; and the bird, of course, was unmistakable. This was a male woodcock engaged in his extravagant courtship dance—and here I was with a ringside seat right on my own front lawn. This was a first-time-ever experience for me, and I can still scarcely credit the excellent fortune that located my house on the very edge of a woodcock's singing field.

Judging from the sound of the calls, this dancing circle must have been at least 100 feet in diameter. Perhaps it was larger. It took the woodcock a long time to complete each circuit, and he must have danced around it five or six times before he again took off—from exactly the same spot, in exactly the same direction—on another rocketing, twittering, spiralling rise into the sky.

Again came a moment of silence, once again a glorious star-burst of shattered silver bubbles, another spectacular plunge to earth, and another strutting, wing-dragging, chest-puffing dance.

Over and over again he took off on that first evening I heard him. Over and over again he repeated his candescent mid-air flurry of bird notes, while the earth grew dark, and the crescent moon turned to silver, and the frogs sang on and on.

Some time during his umpteenth series of beezzping dances the woodcock's voice fell silent—did a woodcock hen come mincing into his magic gyre?—and the evening's rites were ended. I waited for more, sitting there in the night-time chill, but, though the barred owl hooted from the edge of the woods and a gray fox barked in the farther ravine, the woodcock uttered not another beezzp.

For the rest of that enchanted April, every evening was filled with somewhat unpleasant beezzps, with wildly beautiful whirling flights, and with sky-song beyond description. But when April turned into May the incantations ceased and the singing field became just another field growing up in fleabane and daisies, sheltering the nests of sparrows and of redwings, but not of woodcock hens, who prefer a brushier area closer to the marshes.

For the next several years a woodcock appeared each spring

to claim this singing field for his own, adding a flood of enchantment to a season already brimming with magic and wonder. He usually started his rites in March, sometimes early, sometimes late, and he danced through April and early May, and that was the end of it; although one memorable year I heard the first beezzp and saw the first flight on the twenty-sixth day of February, and that particular bird, or another of his kind, kept up his courting until the very end of June!

Now it is not to be thought that these courtship rituals are aimed solely at enticing a compliant woodhen onto the singing (dancing) field. In the first place, there just aren't that many woodhens around, and, in the second place, this isn't the only woodcock male or the only singing field in the countryside. Furthermore, the rites go on with no change in either pace or program (that I can note) through the entire month of April—when the hens are sitting on their mottled eggs—and into May—when they are busily looking after their chicks.

This courtship ritual is a dazzling performance, from wing-dragging strut to spiraling flight to glorious song to spectacular plunge, and I suspect the males keep it up throughout the spring because they become mesmerized by the joy and the excitement of the entire performance.

Roger Tory Peterson says the woodcock's wings make those fantastic twittering sounds during the courtship flights, and I'll accept that as a fact without question, just as I accept, without question, that woodhens are, must be, enticed to the singing fields, though, in all these springs, I've never seen one there. I have observed one thing, though: The courting woodcock on this dancing field always spirals and always dances in a counterclockwise direction.

Last year a woodcock male appeared here only sporadically. This March two woodcocks began using this same singing field but creating separate dancing circles that intersected for almost one-third of their circumferences. The two birds missed one another the first few times around—though I felt they were agitated and that the dancing was not going well—and they each completed one beautiful spiraling flight, one dazzling song, one spectacular return to earth—and then they met, face to face, on their intersecting lines.

There was a long moment of bewildered silence, then both birds gave three fast, stuttering beezzps. They shot up into the air side by side, dark against the sunset sky, then flew off silent and close to the earth, and neither of them has yet returned.

The singing field by my front lawn stands empty and silent and waiting.

The Web
of Warning

For nearly ninety mornings of this long and exciting spring, I have come to Heron Pond before sunup. I have slipped up and over the earth dam in rubber-booted quiet and have hidden myself among the tangles of winterkilled leftovers and new spring growth just above the water's edge.

I have come so often that the two hen mallards are no longer alarmed. They marshal their broods half-way up the pond, and there they preen and feed and go through boot training on the open water, plainly in my sight.

The great blue heron still takes off with a majestic calm while I am only approaching through the marsh, but the three green herons wait until I am atop the dam before they squawk in wild protest and fly off one way or the other across the pond, in an excess of flailings of legs and wings and outstretched scrawny necks. They usually alight no farther away than they were to begin with, and they stand about looking grimly disgruntled in the best heron fashion, but they soon resume their fishing as though I were not here.

The muskrats, though, dive instantly for the underwater entrances to their homes in the banks, and they are gone for the day. It is only with the greatest guile that I ever catch them feeding out of the water or at play, churning up furrows of silver-green water in the gray-green light of the morning.

At my arrival this June day, the earlier frog choirs went silent, but the chorus of several dozen green frogs scattered around the pond and the two bullfrogs on the eastern bank never faltered, but kept up its calling undisturbed. Except, of course, for the two green frogs in the shallow waters that lap the base of the dam, one to my left and the other to my right. These two always stop their sporadic twunking until I am settled into the green tangle above them; then they rejoin the chorus.

The two bullfrogs on the eastern bank are my favorite performers in any amphibian concert. Their voices are rich and melodious, and they sound, a great deal of the time, like steamboat
whistles, like the voices of old stern-wheelers plying the waters of
a distant river. Or, often, they sound exactly like the clear bellowings of bulls from pastures on a farther ridge.

Morning after morning the cast of characters, the play, the
ballet, the concert, goes on at the pond with little change; but on
this particular June morning the two broods of ducklings, with
their mothers, instead of moving slowly up the pond from the dam
area upon my arrival, were already feeding in the upper shallows
close by the mouth of the inlet creek when I arrived.

Two green herons squawked up from their fishing and
flapped their way across the pond as usual, but the third green
heron, who had been perched on the pier, flapped off with a
croaking screech, hit his left wing against a post at the end of the
pier, lost his balance, and toppled into the old wooden boat tied
up there. He scrambled to his feet, peered over the gunwales, saw
me looking down upon him and settled back into the boat, where,
as soon as I moved away, he began a steady tap-tap-tapping of his
sharp, heavy bill against its white-painted sides.

The two bullfrogs on the eastern bank, however, were bellowing in fine voice, and, as soon as I was settled, the two green
frogs close below me rejoined their friends in the pond-side
twunking, while a score of painted turtle heads dotted the surface
of the water as the hard-shelled reptiles came up for air.

While I sat there simmering down to share the world of the
pond, the day brightened until a fan of golden sunlight poured
down the hillside onto the waters of the pond, warming the air as
well as the waters, and casting dark shadows of trees and of
grasses, of rocks and of hummocks, where all had been distinct
and unshadowed before.

Flocks of barn swallows and chimney swifts dipped over the
sunlit water. Bob-whites called from the hillsides, and meadowl-
arks sang from the fenceposts. Redwings o-kra-leed in the marshes, and killdeers complained on flashing wings. And that one
green heron tap-tap-tapped on the inside of the boat, restless at
its mooring.

Now a water snake, long and supple, its tan body marked

with colorful crossbands and blotches of reddish-brown, slithered down the bank beside the pier. It flowed, head first, from the grass into the water, making not a ghost of a splash and sending only the faintest of ripplings into that small corner of the pond.

But every frog in the sunrise chorus went silent on the instant.

Every turtle head went down. The green herons stood at stark attention on their fishing grounds. The mother mallards took their broods up the inlet creek, silently, and on the double. The green heron in the boat ceased his tap-tap-tapping, and the swifts and the barn swallows left the pond. And I felt a real but intangible pressure, a physical constriction, a web of communication I could not apprehend.

The water snake, with graceful undulations of its heavy body, swam out from the bank into the open sunlit waters of the pond. And there it hung, its body swaying slightly in the current, its head resting on the surface, its forked tongue flickering in the air, catching molecules of scent to be read by the special organs in its mouth.

For long, sunny minutes the water snake floated there, reading its own information about the life of the pond and, I'm sure, screening the invisible web of warning that flashed without ceasing from margin to margin and from surface to darkest depths.

The snake swam back to the pier; passed beneath it; half-swam, half-crawled among the overhanging grasses of the water margin, and curled through the water weeds at the base of the dam directly into the territory of the green frog at my right.

Neither the snake nor the frog was visible to me now, and I waited and I watched and I listened. Once the water weeds wiggled briefly and I heard a single, small, suppressed cry. . . .

There followed not a motion, not a sound. And the tension around the pond did not relax.

I sat in silence for another thirty minutes, waiting for the snake to reappear, its body bulging with the swallowed body of the green frog; but I waited in vain. I did not see the snake, or the frog, again. Nor did I, any longer, feel the pressure of the pond world's fear.

Although every frog around the margins remained in deepest silence, and the mallards did not return to the pond that morning, after a while, turtle heads began to pop up like small dark ping pong balls over the surface of the water, the two green herons resumed their fishing, and the green heron in the gently rocking boat renewed his vigorous tapping.

I rose quietly, crept to the pier, and looked down into the boat. Dozens upon hundreds of small, dark flies were clustered into the sun-warmed protection of the boat's curved sides and the green heron was methodically picking them off with the sharp point of his bill, his crest rising and falling at crazy angles as he worked.

The green heron looked up, belatedly, and saw me, so closely by, peering down upon him. With one wholly terrified squawk he shattered the silence, and with one wild flap of his frantic wings he scattered all the flies as he tumbled over the gunwales into the water and took off haphazardly for the haven of the opposite shore.

Once again a taut web of alarm held the pond immobile. But this time, perhaps because I was up and stirring and because the warning was given in a squawking voice that even I could hear, the web seemed neither so mysterious nor so subtly pressing—upon me. I could not, in fact, feel it at all except that I was aware of it in a remote, third-person, kind of way.

But I knew it was acutely first-person to every other creature at and in the pond. And I knew who was the cause, this time, of their gripping fear. So I went, slinking stealthily, up and over the dam and out of the blossoming marshes. . . .

What Singeth the Cardinal?

My dooryard cardinal started out his singing career this year, on January 24, with a long, soulfully whistled rendition of "Peek-a-boo! Peek-a-boo! Peek-a-boo! Peek-a-boo!"

His red neighbors on all sides, apparently intrigued with this highly original cardinal phraseology, took up the song and, from then until the Big Snow of February, my semiwild hermitage rang with the sounds of a tip-top peek-a-boo game that was all noise and no dodging.

But the February storm imposed a moratorium on singing, and by the time songs again became the fashion, the cardinals had forgotten their new lines and went back to an old favorite in this area: "For years and years and years and years. . . ." sung, sometimes, in tones of deep devotion, on other occasions with undertones of sadness or of disillusionment.

Of course, authorities say, no bird is capable of making a consonant sound. These interpretations of their songs which we so freely make are the results of hearing with human ears—ears strongly accustomed to hearing consonants. As for attributing human emotions to those same songs, tht is the sin of Anthropomorphism, with, as you will note, a capital A.

I submit to the authority of these authorities, but what, I wonder, am I to make of yesterday's incident when a male cardinal sang a most sincere-sounding, "For years and years and years and years . . ." and then added a swinging, "Eh wot, boys?" He sang it three times. I couldn't be mistaken.

As I write this, two cardinals are singing, one on the edge of the woods, one in the apple orchard. The first cardinal is musically whistling, "What cheer! Quoit! Quoit! Quoit! Quoit! What cheer!" The second is, in his turn, tunefully responding, "What cheer! What cheer!" and then, sharply and unmusically, "Quick! Quick! Quick! Quick!"

The "What cheer!" song is standard cardinal from Florida to southern New England. "What cheer! What cheer! Birdie, Birdie, Birdie, Birdie! What cheer!" is the correct rendition, but this is the first time I have consciously heard even a single "What cheer!" in the fifteen years I've lived in this sheltered enclave.

Cardinals are homebodies, year-round residents who don't migrate any further than the nearest feeding station and, like their human counterparts who become isolated from the mainstream, they tend to develop their own local dialects and their own local patterns of song.

But a cardinal voice is a cardinal voice. Whether he is singing in Georgia or in Connecticut, or whatever may be the pattern of his song, the voice is unmistakably cardinal, even to the ear of the rankest amateur.

I have been referring to the male cardinal as though he does all the singing for his family, but this is not so. The female cardinal is a singer, too, and she sings, so far as I know, the very same songs as the male.

Some writers say she sings in a softer, more subdued voice than does the male, and this is probably true, generally, but I have heard, and seen, female cardinals pouring forth their songs in as loud and uninhibited a fashion as ever I've seen and heard a male perform.

These same writers say she sings only in the mating season and with this, I think, I do concur. I cannot remember that I have ever seen and heard—it is necessary to do both at the same time —a female cardinal singing except in spring and early summer.

The male sings for most of the year, quieting in late autumn and starting up again in January and February. This extended period of beautiful singing almost brought an end to the cardinals of America. During the latter years, at least, of the 1800s and the early third of this century, the brilliant males were captured by the hundreds and shipped as cage birds to Europe where they were known as Virginia Nightingales. Their colors quickly faded in captivity, but in their loneliness they sang almost without ceasing and so they were in great demand.

Lest there be anyone who does not know: A male cardinal is a brilliant, all-over-red bird, though his back is sometimes grayish or brownish in the fall and early winter. His crest is tall and prominent. He has a black patch around his bill—it just touches his eyes and darkens his throat—and his orange-red bill is thick and heavy.

He is the redbird of the southland. His red is the red of the cardinals of Rome. So red is he that Linnaeus named him *Cardinalis cardinalis cardinalis,* although he is now known officially as *Richmondena cardinalis.*

The female cardinal, and there are those of us who think her prettier than her mate, is yellowish-brown with orange-red in her crest, her wings, and her tail. Her orange-red bill is, like his, thick and heavy.

These birds pick up sunflower seeds and pumpkin seeds in those thick and heavy bills, expertly slice away the coverings and, with the aid of the slender tongue, deftly maneuver the kernels into the mouth with rarely ever a slip in the procedure. That's a neat trick—no hands—and they eat with the preoccupied air of someone at dinner table trying to locate a fishbone in his mouth.

Cardinals are complacent, self-sufficient birds not easily rattled. They are ready to do battle if necessary—very ready to chase an intruder of their own species from home territory—but they are never in the vanguard of the alarmists.

The young birds pair off at the end of their first winter—definitely in their first spring—and they remain mated for life. They fly together, feed together, roost together all the year around. But each spring is a new courtship, a new wooing, when the male pays extravagant attention to his mate, calls her to special foods, and feeds her tidbits beak to beak.

They build their nest of twigs and grasses and leaves, fashioning a rather deep cup and lining it with rootlets and fine grasses —and hair if she can find it. It isn't the neatest nest in the world and, usually, they place it rather low, but they hide it fairly well in the twisted tangles of bushes and vines and briars.

Actually the female does most of the building, but the male helps her by following her in and out, by watching her closely, by singing to her hour by hour, by feeding her an occasional special seed.

Brilliantly colored male birds of any species do not usually incubate the eggs and apparently the male cardinal does not, but he goes in and out of the nesting place so often to feed his mate or just to be near that he must surely draw as much, or more, attention to the site than if he were actually brooding the eggs.

The three or four eggs she lays are small, white or greenish-white, and variously blotched in shades of brown. A characteristic of a cardinal's clutch is that, very often, one egg will be markedly different from the others, perhaps much smaller and/or with a color change.

When the baby birds hatch, the male bird works just as hard as the female at feeding their young family. When she begins to lay her eggs for their second brood he takes over complete care of the growing, scattered and still-demanding little ones from the first nesting.

By the time he has these youngsters fully fledged and off on their own uncertain wings, the second brood is ready for round-the-clock attention.

Rarely, but not usually, a third brood is raised. Usually, with the fledging of the second brood, the parents call a halt to reproduction and go back to their own song-filled and companionably platonic lives.

This season's youngsters will probably stay close to their parents' territory through the winter and disperse, in the spring, into their own properties as nearby as population pressures will permit—all of them whistling their provincial songs, with variations, in the best of their beautiful all-cardinal voices.

Cricket Time

The chorus of bird songs around my home these August mornings has dwindled to a few rollicking calls from the Carolina wrens, occasional faint-hearted flutings from the wood thrushes, the idiot chuckle of a cuckoo in the distance and notably unenthusiastic chirrups from the robins in the orchard.

The birds are tired out with their months of relentless activity. Many of them are feeding their third, or even their fourth, brood of young ones, and most of them are dispirited with the summer heat and half-ill with the discomforts of molting. Gone is all the sweetness and joy of springtime's mating songs, the splendor and dash of territorial proclamations, and no one bursts into spontaneous carols just because a new day is dawning and the morning sun is about to appear.

During August's hovering summer, the mating calls and the vocal claims to territory are transferred to the other end of the day. As evening's twilight deepens to dusk and then to darkness, thousands of creatures with six legs and no feathers announce to the world at large, and to their own species in particular, that they have laid claim to a niche, that they are ready to mate, and, just possibly, that they are bursting with health, good spirits and convivial song.

The individual sounds made by most insects are not melodious to the human ear, and there are August evenings when the simultaneous rising of all their noises produces a roaring so intense and so discordant that only the most determined of nature lovers can appreciate its subtleties or bother to sort out its components.

But to isolate a component, to sort out an individual song, is one thing. To locate the singer is quite another. The sounds are ventriloquistic to begin with, and they are tossed and bent and redirected by leaves, by walls, by twigs, by grasses, by any change in the insect's position—and the human ear is far more directional and more easily confused than you may think.

In the ringing darkness of an August evening, I walked down the slope of the yard to put out a soupçon of table scraps for little forest wildlings. I was aware of the clangor but I was only casually listening. My ears sorted out the songs I knew and indifferently tossed aside the sometimes tantalizing unknowns.

I lowered my head to pass beneath a dogwood tree at the edge of the woods, and a ringing trill I had pursued in vain for years suddenly shrieked in my left ear, loud, piercing, painful. I snapped on my flashlight and there in its beam, in the curl of a dogwood leaf, a strange, thin insect, all purplish-red and slightly less than an inch in length, balanced itself on long grasshopper legs and mistily vibrated the gauzy wings it held almost perpendicular above its narrow back.

The shrill musical trill coming from those vibrating wings would carry for a quarter of a mile, I'm sure, but at that moment, a purplish-red replica of this wing-singing fellow came creeping just from the furthest reaches of this same branch of the dogwood tree.

Her body was even thinner than his and she kept her wings wrapped tightly about her. She moved purposefully toward the singer, yet her manner was secretive, even furtive, as though she came to steal and not upon invitation.

She came creeping up behind him on her too-long grasshopper legs, creeping slowly and silently, and when she had perilously climbed the thin stem and stood upon his leaf, she paused and cleaned her long antennae, drawing each one carefully through her mouth parts. I don't know whether or not he knew she was there. He did not look around nor pause in his trilling.

She tapped him with her newly cleaned antennae, then climbed upon his back and began to nibble at something near the base of his wings. He stopped his singing then, although his wings remained upright, and carrying her, he moved out of my sight to the underside of the leaf.

Shamelessly, I bent my head and looked beneath the leaf to see his abdomen curve upward to her while she continued to feed upon whatever lay at the base of his wings.

By the purest of serendipitous encounters I had found one of the nighttime singers that had eluded me for years. By even greater serendipity I had witnessed their mating. Now I turned confidently to my field guides to round out the experience by discovering the names of the creatures I'd happened upon.

They had grasshopper legs and antennae half again as long as their bodies, so I looked for them among the grasshoppers and the crickets. Most of my field guides showed the pale-green or white snowy tree cricket which chirps but does not trill. Some showed, also, a greenish tree cricket that trilled. But not one mentioned, even in a footnote, a trilling tree cricket in purplish-red.

Because their bodies so closely resembled the pictured snowys, I was certain that my pair were tree crickets—and tree crickets they turned out to be. Extended reading, far afield from the guides, turned up one casual reference to trilling tree crickets that are "crimson-red in color." My pair had been more purplish than crimson, but I've since found trilling tree crickets in hues from purple to lilac and from crimson to pink, so color variations are apparently in order.

So far as I can discover, the wing-music made by all male tree crickets is pleasant to human ears—unless the ears happen to be too close to the cricket. But that same statement cannot be made for the songs of the grasshoppers, the Katydids, and several other members of the order Orthoptera.

Yet, strangely enough, while female grasshoppers and Katydids have genuine ears in their front legs and can appreciate the rasping wing-songs of their males, the female tree cricket, so far as we can discover, has no ears of any kind, and all the real music of the male tree crickets goes unheard by the very ones it should be intended to attract and please.

Female tree crickets are actually attracted to their mates by the odor of the sweet and smelly liquid exuded by a pair of glands at the bases of the males' wings. These glands are exposed when the wings are raised, and I suspect, though I've never read it anywhere, that the vibrating of the wings may cause or increase the flow of the liquid or help to waft its fragrance into the darkling air. The trilling music that results may be entirely coincidental.

Almost immediately after mating, the female tree cricket lays her eggs. With her sharp ovipositor she punctures twigs and stems and deposits her eggs in the soft pith within, usually killing the plant beyond that point. She seems to prefer raspberry stems above all others.

Tiny tree crickets hatch from these eggs the following spring and they immediately begin to feed upon aphids as well as upon leaves. They may molt eight or ten times before deep summer, when their wings appear and they are ready to add their music, coincidentally or not, to the great late-summer chorus of the insect world.

The Heron Casts
No Shadow

The great blue heron has been standing in the shallows on the southwestern border of the pond for more than an hour now, and I have been lying close to the roots of the goldenrod and the asters on the earth dam, watching him, for exactly the same length of time.

I don't, in fact, know how long he has been standing in that selfsame spot, because he was there when I arrived, sliding on my stomach over the crest of the dam and slithering down through all the September flowers so that I could peer out at him from the level of the duckweed on the water's edge.

He is still a good 300 feet away from me, but I dare not attempt a closer crawl, for he is a jittery bird who takes off for other waters when he sees me approaching a quarter of a mile away.

His predecessor, who had fished this pond for years, was of a calmer nature. He could fish undisturbed while I sat in plain sight at lesser distances than this, or, if I attempted a watch he considered too intrusive, he'd flap to the top of the old oak snag across the pond and sit there preening his feathers, while my dog and I walked around the pond and considered him from every advantage—even from overhead if I climbed the steep pasture-ridge behind him. He studied us as we studied him, but he did enforce a sixty-foot individual-space allowance from which we could commune in mutual respect and confidence.

We learned a lot from each other, that old heron and I, but this new bird wants half a mile between him and any prying human being and he abandons the pond with an exaggerated air of personal insult and invaded privacy while I'm still far off on the horizon and haven't yet set foot in his watery domain.

My getting this close to him this morning was a matter of accidental timing, of his being otherwise occupied at the crucial

moments when my distant advance was, inescapably, visible to him—for I don't think he has suddenly decided to accept me as part of the pond's biota and I certainly did not know he was here.

I made my creepy-crawly approach over the dam just on the off-chance that he might be fishing or frogging or meadow-mousing along the weedy borders opposite to the rising sun. And sure enough, there he was, up in the shallows by the sand bar and so nearly invisible in the blue-gray mists of the morning that it took me more than a minute to locate him.

This always puzzles me. How can he possibly be so indiscernible? A mature great blue heron is no feathery bit of fluff. He's feathery, yes, and rubbery-necked, but he stands in plain sight, out there in the water, tilted up on the thin stilts of his featherless legs, and his body measures more than four feet from his head to his tail. He is large; he is dark; he is solidly oval against the thin vertical lines of the pondside vegetation, yet he merges; he blends; he disappears.

They even say, the mythmakers, that a great blue heron never casts a shadow. But this is only partially true, for herons are not immaterial spirits transparent to the light of the daytime sun. They are real, tangible, organically material birds who give rise to myths and occult speculations because, aside from the two or three months each spring when they gather into their tribal rookeries to mate, to nest and to rear their young, they are apt to stalk the shallows of distant ponds, silent and solitary, yellow-eyed, mysterious and inscrutable.

But not shadowless. Those particular myths arose because, as a general rule, the shadows that herons cast fall upon the shores and not upon the waters where they fish. They arrange themselves so.

In the morning light, this blue heron, if he is feeding at this pond, fishes, usually along its western edges where the shadow of his body falls, with careful intent, only into the bur reeds and the sedges. There its gray darkness disrupts none of the water patterns below him but it does frighten an occasional bank-sitting frog into making a frenzied leap toward the water, where it is instantly speared and swallowed by the breakfast-hunting heron.

He hunts the irregular coves of the pond's southern edges in the summertime and of its northern edges in the winter, hunts its eastern borders in the afternoons, fishes from the protection of overhanging grasses and willow clumps at any time of day, and prowls the shallows of every location in fogs and drizzles and under shadow-erasing cloud covers. Thus does our great blue heron never cast a shadow.

Neither does he, in his hunting and fishing, ever let the water drip from his uplifted foot to cause fish-alarming ripples on the top of the water and to send the crayfish and the salamanders scurrying for cover. I'm not at all sure how he does, or doesn't do, this, but I think it has to do with the way he bends his foot backward as he lifts it from the water. In this position his foot hangs diagonally downward, but is actually upside down, and his claws form concave pockets which must catch and hold the droplets that run down his polished toes.

Every movement, when he is hunting, is made with such consummate grace, with such imperceptible muscular change, that our human eyes rarely see his actual motions. We only see that the head has turned, the neck has stretched, the step was taken.

Imperceptible and in exceedingly slow motion most of his actions seem, and yet the powerful slashing downthrust of the neck-driven beak into the soft body of fish or frog or snake or mouse is so fantastically fast that the most intent human eye sees only the end result and not the elegance and vigor of the strike itself.

Once I sat on a catwalk in a swamp in south-central Florida listening to the gurglings and gruntings and bellowings of the swamp's unseen inhabitants and watching the faint, beginning light of a new day slowly penetrate the hanging mists and mosses about me, when a great blue heron walked in formless stealth into the wide pond of black water over which I sat guard.

He was barely discernible in the dispersing darkness and he fished over the entire pond, up the middle and from side to side, moving in precision and in absolute silence, though I suspected that the din of the swamp-dwellers all about us might serve as cover for any sound he made. But as the sun rose, the cacophony shifted from the waters and the banks of the waters to the trees and the bushes around about and the heron walked in the telescoped rays of the sun and then in the full ambience of the light of day, and he walked without a splash and he walked without a ripple.

At first it was his proximity and then his size that enthralled me. He was not ten feet away, his blue-gray back on a level with the slatted floor where I sat. And he was taller than I, his body as large as my own. Then I became engrossed in watching how precisely he balanced himself as he walked, planting his feet firmly in position and only then bringing his body forward to catch up. Analyzed, it seemed an awkward, disjointed walk, yet it was smooth-flowing and counterpoised, and allowed him a steady view into the water should his footsteps disturb an edible creature from the bottom mud.

But what could he see of animal movement in that foot-deep water, I wondered, when its dark surface was almost completely covered by a floating curtain of dainty, pale-green duckweed.

It was then I noticed that his emerging foot swung backward and upside down the instant it cleared the water and that, in descending, it came right side up at the very moment he reinserted it—as though through a slit between the body of the water and its surface skin—sliding it diagonally downward beneath the floating plants, down to firm footing on the bottom, causing not a ripple, making little disturbance in the patterns of the floating duckweed, and, as far as I could see, freeing no upward-rising column of finely spiraling mud.

Suddenly, with a quick, powerful downward thrust, his lazily S-curved neck became a straightened line, and, when he lifted his head from the water, a flattish fish quivered violently on the fierce sword-point of his formidable yellow bill. With a slight toss of his head he flipped the fish free, caught it in his open bill, and swallowed it, headfirst, protecting his tender gullet from the bony tips of its trembling fins.

The fish could scarcely have reached the bird's crop when he snapped up a big green frog in his open bill. The frog's long, spotted legs kicked the air in frenzied desperation and his thin, froggy death-scream reached my ears across the narrow space of water. . . .

Yet the essential presence of the great bird spoke not of terror, nor of force, nor of secret, swamp-hidden rites and mysteries. He explicated, rather, a practical and pragmatic thereness; a great-bodied creature filling his niche with skilful competence, without dramatics and without fanfare, giving no quarter and certainly asking none, dominant, secure and unquestioning.

The great blue heron of the Florida swamp bent his knees, hunkered himself beneath the catwalk and hunted his slow way off into the watery mazes of the shadowy, moss-hung wilderness. But the great blue heron of my neighbor's pasture-pond hunted and fished his careful ambling way down the western bank of the pond until he stood, neck folded, body tilted, gazing into the still waters of a cove thirty feet from me across the corner of the pond.

He froze a sudden downward slash of his head in mid-thrust and stared instead into my flowery cover. I had not moved. I was so relaxed that I was scarcely breathing. Yet I had been discovered and my little adventure was at an end. Perhaps he saw me. Perhaps he purely sensed my alien being-there.

Offended and righteously irate, he sought to quit the scene, but in his kingly haste he lost every vestige of his kingly dignity. His tube neck gangled at full-stretched length, his stick legs hung angularly askew, his powerful wings flailed ineffectually at the motionless air while he desperately fought his ungainly way aloft.

But once fairly launched, he flew with dignity and strength. He scooped great gobs of supportive air under his deep-curved wings, eased his legs up against his soft underbody and allowed his feet to trail behind. In regal mastery, he relaxed his neck into its accustomed S-curved position, rested his head comfortably back between his shoulders and solemnly flapped away to more hidden waters where he could fish in royal, solitary and un-spied-upon seclusion.

An Afternoon
of Monarchs

Since two o'clock this afternoon, monarch butterflies in increasing numbers have been fluttering back and forth along the front of the woods by the side of my house.

At first there were only a few of them, and I might not have noticed, except for the way they flew at the woods and away, at the woods and away, as though the trees might have disappeared while their backs were turned or a passage-way have opened through.

To right and to left, to east and to west, they fluttered. Four hundred, five hundred feet each way, testing the woods as they went, but never flying quite far enough along its edge to reach the roadway and the open fields that stretch, clear and inviting, to the south-southwest on the butterflies' preordained route.

As the warm and sunny afternoon hours went by, more and more monarchs joined the apparently stymied migrants before my woods until half a thousand butterflies fluttered in a long uneven line of orange-and-black wings against the sunlit purples and russets and golds of October leaves.

I say the "apparently" stymied migrants because I'm not sure that they were really baffled. I have seen monarchs behave this way in front of my woods in other Octobers, and I've also seen monarchs flying right over the top of Mt. Mitchell (6,684 ft.), the highest peak in the east, so I scarcely think they were balked by the height of my white oaks.

I realize, of course, that the butterflies at Mt. Mitchell were flying only a few feet above the ground, not six thousand feet, or even a hundred feet, above it, and that the sudden height of my woods might be, at this moment, an actual barrier to their southwestward progress; but these butterflies had all the appearance of deliberately stalling.

They flew to within fifty or even twenty feet of the open

passage of the roadway and fields, then turned and fluttered back in the opposite direction. Every once in a while fifteen or twenty butterflies slipped into the corridor where the trees are kept low beneath the electric lines, but they flew only a little way along it and then come drifting back.

This northern front of my woods appears to be a regular camping ground for the monarchs, whose ancestral flyway passes this way. These particular butterflies may have arrived too early this afternoon to settle down and too late to fly on to the next recognized overnight spot. Probably they did not *know* this. Perhaps they were truly searching for a passageway through the woods but their timetable did not allow them to find it. Perhaps they were simulating a search to take up the time or to keep their flying muscles limber. Or perhaps they were doing nothing of the kind. . . .

All of these butterflies had just fed richly and well on the nectar in the flourishing late goldenrods and asters in the fields and marshes beyond my house. Through all the morning hours they drifted, feasting, from flower to flower, not, apparently, intent on getting anywhere until their bodies were heavy with sweetness. Then they left the low fields and found the sudden height of my woods a barrier to their southbound flight.

Now the sun is almost touching the western hills. A noticeable chill is creeping into the air and the butterflies in their regal coloring are settling into camp. They are bedding down in small groups on the branches of the frontline trees, mostly on the low purpled dogwoods, where they will spend the cool night in a nearly torpid state.

They won't get an early start in the morning, either, for the woods stretch away to the east and the warming rays of the sun are late in penetrating to these low branches on the northern front.

But, as the sun climbs the sky through the morning hours, the butterflies will slowly fan their wings in the warming air, spread open and close, spread open and close. They will shake off their torpor, loose their holds on the branches and drop on outspread wings. Each one will flutter in an orienting circle and then, in a long orange-and-black procession, they will fly directly to the open passageway of the road and the fields beyond, and head off without a pause to the south-southwest.

They do not fly together as a migrating group. There seems to be no social organization, and no butterfly is in charge of the flight. They are all simply going to the same place at the same time, like summer vacationists thronging the roadways to a favorite resort.

These monarchs do spend the nights of migration and the long months of their overwintering hanging together in somewhat separated clusters, but whether their clusters are made up of groups that flew south together, or even if individuals remain in the same clusters all winter, I do not know.

Tomorrow afternoon, or the next, there may be another flight of over-camping monarch butterflies at the edge of my woods— or there may not. Some years there are flights every day for a week, and some years I miss them entirely. But the wonder is that the flights occur at all. And the marvel is that they are repeated in detail year after year.

For the monarchs flying south this fall are not the same monarchs that flew north last spring. They are not even the children of those monarchs. At the closest possible relationship they are their great-great-grandchildren, and they may even be further removed than that.

The migration behavior of the monarch butterflies is neither taught, nor learned, nor passed on to the next generation, or to the next, but is held in trust for a distant generation. And what triggers the response we do not know.

The urgency that sends this last generation of the summer's hatchlings fluttering off southwestward may be a response to the waning length of day, to the lowering temperatures of the nights, or to the aging of the milkweeds on which they feed as larvae. Whatever the cause, the Monarchs that emerge from their chrysalids in the fall are driven by an urgency to migrate instead of the urgency to mate that impelled their recent ancestors.

Some unknown circumstance causes a whole generation of butterflies to migrate after several generations have stayed at home and flourished.

And something even more unknown causes this season's mi-

grants to follow the exact routes their great-great-grandparents
took a whole year ago, and to repeat, exactly, some of their idi-
osyncracies that are unexplainable in the first place—such as making precisely angled turns above certain wide bodies of water, or
expending whole sunny afternoons fluttering uselessly before the
russet and gold and purple of my woods.

Lead-Backed Red-Backed Salamanders

The birdbath in my backyard is a shallow earthenware basin that sits in a hollow made by the pressure of its own weight in the soft earth of the bird-feeding station beside the lilacs.

When I tip the basin to empty it, clean it, and add fresh water, I usually find under it such things as pill bugs, centipedes, earthworms, and slugs; but on this lead-gray November morning I found two salamanders lying beneath the basin, each one curled into a smoothly broken cavity in the earth, just the size of its own small brown-black body.

I smiled to see them because they are the first salamanders I've seen this fall and because they make this the thirteenth successive year that I have found members of this species settling into hibernation somewhere along the edge of my woods.

They are red-backed salamanders, probably the most common terrestrial species in the area, but these two belong to the lead-backed phase of the red-backed salamanders. Lead-backs are uniformly dark gray to black instead of having the wide red, orange, yellow, or even light gray stripe from head to tail which marks the red-backed phase.

I consider the lead-backs in my area more brownish-black than the dark-gray-to-black the field guides stipulate. However, all of them have undersides mottled in black-and-white which gives the salt-and-pepper effect that, the same field guides say, is the outstanding identifying characteristic of red-backed salamanders, whatever color their backs may be.

As a matter of fact, I have never found a *red*-backed one. Nor a gray- nor a yellow-backed one either. All I've ever seen have been deeply and darkly, and often brownly, black. And I rarely catch sight of any of these except under the birdbath where the resident crawlers are favorite foods.

These two under the birdbath this morning were in a torpid

state, but not yet, I hope, completely hibernating. They need to go much deeper into the earth to escape death from freezing. If the sun breaks through the clouds this afternoon, it may warm their cold blood so they can continue pushing their way downward. Perhaps even the warmth of the water above them this morning will rouse them enough.

Most terrestial salamanders hibernate hither and yon, one here, one there, but red-backed salamanders are usually found two by two, or in clusters of small groups, not curled up together, not touching one another, just in fairly close proximity under the same log or in the same pile of wood chips.

And this is for a reason. Most kinds of salamanders assemble in numbers for romantic purposes in late or early spring and then wander, singly, farther and farther apart during the summer months. Red-backed salamanders, on the other hand, gather for breeding during the fall and then go into hibernation while they are still more or less in assembly, soon after their courtships are completed.

And salamanders really do go a-courting, whether, like most, they choose the soft gray nights of spring or, like the red-backs, they prefer the mellow nights of autumn. Salamanders make no music to call their multitudes together as do their fellow amphibi-ans, the toads and the frogs, because salamanders have no vocal cords to sing with and no ears to hear a voice if one were raised. But springtime-mating salamanders frolic through great carnivals of courtship without the harmonics of song. They waft their per-fumes upon the air, they choose their partners, and they dance in the waters and they dance upon the earth.

(Male salamanders manufacture their perfumes in special and aptly named hedonic glands located under their chins and at the bases of their tails. Female salamanders have no scent glands that we can recognize as such, but the moist secretions of their skins contain the fragrances that identify them to the males both as to their species and as to their sex.)

But the courtship of the red-backed salamanders is a quiet and private affair conducted without the aid of a carnival atmo-sphere.

On a warm night in autumn, a red-backed male salamander running about the woodland floor is attracted by a red-backed female's very proper and correctly titillating fragrance. He runs quickly to her and begins circling about her, rubbing his perfumed chin on her head and brushing her body with his scented tail.

When the various perfumes and the getting-acquainted ceremony have worked their magic, the two salamanders swing into a courtship dance that begins with much chasing and pirouetting and ends in a tight gyration with her head nestling on the base of his tail and his head lying on the base of hers.

At this point the male places a small gelatinous cone, called a spermataphore, on the earth, and the female, passing over it, picks it up in her cloaca and stows it away in a special receptacle, the spermatheca. This is the climactic moment. Now the dance slows down and dwindles to an end. The courtship is over, mating has been achieved, and the whole summer's gathering of salamander energy is justified.

As daybreak nears the two creep under a log or a stone close by their trysting place to sleep through the hours of light. As long as the weather stays warm they will remain night-time wanderers, more or less together, in their woodland territory. But, as the temperature drops, they will go into hibernation, resting lightly at first, and then sinking slowly into torpor.

It is easy for red-backed salamanders to crawl under things. Their lower jaws are immovable and this makes the front ends of their bodies stiff enough to force their ways into narrow crevices.

They *should* pass the winter below frostline, but a great many of them don't dig deeply enough, and, while some are known to come successfully through the cold months even though they've been only three or four inches below the earth's surface, the wintertime mortality rate for red-backed salamanders is thought to be pretty high.

If all goes well with this theoretical pair of lead-backed red-backed salamanders we are watching, they will come out of hibernation when the weather warms in the spring and each will then go its separate way. The female will not look for a place to lay her eggs until May or June, or, even, sometimes, not until July.

When her eggs are developed within her body, she finds a small cavity in a fallen log or in the earth beneath a log, and there she deposits her colorless eggs. As the eggs move from her body toward the outer world, the spermataphore, which she picked up and stored during her courtship dance many months before, is ruptured, and, only then, are her eggs fertilized.

She lays only a few eggs—from three to thirteen in all—and each egg is about one-fifth of an inch (5 mm) in diameter. Each egg is protected by two jelly envelopes, and all the eggs are suspended together, like a cluster of colorless grapes, from the top of the nesting cavity.

It takes from one to two months for these eggs to hatch, and during all this time the female red-backed salamander remains with her eggs. She is not so much guarding them from predators as she is protecting them from dehydration. The same glandular secretions which keep her own skin constantly moist also dampen her eggs as she lies coiled around them or creeps in and out among the hanging cluster.

When the young red-backed salamanders come out of their eggs they will be fully formed, able to run about, to capture insect food, and to crawl into cover just as their parents do.

They will have gills, though—fancy, deeply-cut, "staghorn" gills—that mark them as salamanders in the larval stage. But the gills were larger while they were still in the egg. Already they are being absorbed into the little bodies. In only a few days even the gills are gone and the larvae look exactly like their parents except that they are scarcely three-quarters of an inch long.

It will be two or three more years before these young salamanders reach adulthood and are ready to take their rightful places in the long line of the lead-backed red-backed salamander generations.

How Doth the Skunk Cabbage?

This morning I stood on the edge of the marsh where the great gray elephant-rock is slowly sinking into the black mire beneath it and surveyed with silent respect the thousand or so purple-green-and-yellow teepees of this great springtime encampment of the Skunk Cabbage tribe.

In childhood I *knew* that families of tiny Indians lived in these teepees. I *knew* that they gathered here from all over and that the place fairly rang with all the noise and bustle and to-do of camp-ground activity. I could feel the excitement and the busyness. That I could neither see it nor hear it made no difference. I *knew* that something not ordinary was going on out there.

Today I know a great many more scientifically established facts about the life processes of *Symplocos foetidus* than I did away back then. But, in spite of those documented facts, I had the same perception, this morning, of stir and motion and barely-contained excitement among the skunk cabbage teepees and the feeling that whether the workaday activity of the skunk cabbages is being carried on by tiny Indians or by cell division, osmosis, capillary action or whatever, there is, very definitely, something exciting going on out there.

Perhaps the tiny Indian theory is as easy to handle as the scientifically established facts. There really is, for instance, the equivalent of a small campfire burning in the center of each small wigwam and it raises the temperature within the skunk cabbage spathe to near tropical comfort.

Biologists have long known about that heat. The theory was that skunk cabbage spathes (the wigwams or teepees of first spring growth) generate a heat that keeps their temperatures twenty to thirty degrees (Fahrenheit) above their surroundings and melts the ice and the snow around them to distances of one or two inches, thus making wintertime growth a possibility.

It was generally thought that this heat was turned out in the process of growth itself since all growth produces heat. It appeared that the heavy skunk cabbage spathe just happened to produce more heat in its growing than the average plant does.

But, just recently, Roger M. Knutson, professor of biology at Luther College in Decoran, Iowa, with thermometer in hand, has shown that it is the spadix, the knobbed club within the spathe, which generates heat in prodigious amounts, and that this rising temperature is directly concerned with the blossoming and the fertilization of the skunk cabbage flowers. This is a startling discovery because such physical phenomena have heretofore been associated only with the animal kingdom.

Mr. Knutson discovered that the spadix can generate heat from thirty-six to sixty-three degrees above outside temperatures and that it can hold these temperatures for two weeks or longer. While the outside temperature remains even just a little above freezing the spadix will keep the temperature inside the spathe at a fairly constant seventy-two to seventy-four degrees. It cannot reduce temperatures, however. Prof. Knutson found that temperatures inside the spathes in sunlight in the month of March were usually near eighty degrees.

Every person who has ever stumbled over a field guide knows that skunk cabbage is the earliest of our spring flowers to bloom. And many people, out looking, have thought that the spadix itself is the acceptably bizarre flowering of so strange a plant as the skunk cabbage. Few people, I suspect, have ever seen the actual blossoms with their lavender, tissue-paper petals (or sepals) spread out flat upon the rough surface of the knobbed top of the spadix.

The tissue-paper flowers are real and perfect, possessing both pistil and stamens, and the spread stamens lie flat across the flowers, covering them with a fine netting of straw-colored filaments.

The flowers, however, are not self-pollinating. Tiny swarms of tiny flies, feeding and courting and mating within the dusky confines of the warmly heated wigwam, apparently pollinate the skunk cabbage flowers in the normal course of their own feeding and courting antics.

But where, with late-winter temperatures hovering near the

freezing mark, do these swarms of tiny flies come from? No one, I think, knows for sure, but it seems likely that the last generation of these flies, in the warmth of the autumn before, laid its eggs on the black, wet earth close to the dying old skunk cabbage leaves.

Then, when the heat generated by the new, growing spathe melts the snow and ice around it, the embryos in those eggs that happen to be lying close enough are stimulated to develop (responding to heat as well as to increasing hours of daylight) and they hatch as the spathe opens and the spadix generates its welcome heat in the murky enclosure.

In line with nature's frugality and abhorrence of a vacuum, there is, inevitably, a species of small spider that flings its filmy web across the open door or from knob to floor of practically every skunk cabbage spathe in the marshes.

So far as I know, the presence of these spiders neither advances nor deters the life of the skunk cabbage. They are there for their own advantage, enjoying the unseasonable warmth inside the skunk cabbage spathe, harvesting unwary individuals from the cloistered crop of flies, and insuring their own progeny by engaging in hazardous spider courtships.

But, to get back to the skunk cabbage: Each single pollinated flower produces one large, dark, lima-bean-shaped or pebble-shaped seed that develops directly under the skin of the spadix knob. In the fall the spadix rots away and the dark seeds fall upon the ground. Each seed then puts down, directly beneath it, the tiniest beginning of what is to be its permanent, long-lived, underground stem, and spreads out a circle of tiny, thin, crinkly roots.

Those thin and crinkly roots spread out horizontally and then grow down into the earth. At some unknown signal all of them, together, contract upon themselves, wrinkling, drawing back, and they pull the seed downward an infinitesimal distance.

By this action the seed is eventually buried, but the process is so slow that it takes two and sometimes three years to sink the seed below the surface of the black marsh mud. Five to seven years will elapse before that seed sends up its first small spathe.

Beneath every skunk cabbage plant an underground stem plunges like a tap root straight down into the mire. Hundreds of true roots, pencil-sized, wrinkled, and ridged, extend horizontally, in circles, reaching out a foot or more beyond the drip-line of the plant.

Each year's new roots grow out in a circle from the top of the underground stem, directly beneath the plant. They reach out, horizontally, perhaps eighteen inches or more, anchoring as firmly as they can into the watery mud, and sometime during the season, again, at an unknown signal, they all contract at once, wrinkling themselves, shortening themselves, and they pull the plant downward into the marsh perhaps as much as one-eighth of an inch.

The underground stem lengthens accordingly. And it grows thicker. Some underground stems are a foot long and several inches in diameter. Some are two feet long. Some of the oldest may be even longer and very thick.

No one knows how old a skunk cabbage plant may be. Some of those plants in the bowl-shaped marsh where I stood this morning may have been growing right there when Columbus landed at San Salvador. Some of them may be older than the redwoods, or the bur oaks, or the bristlecones.

I cannot feel that my senses misled me, either in my childhood or on this very morning. Little Indians or no little Indians, something, unmistakably, is going on out there.

The World's Most Valuable Burier

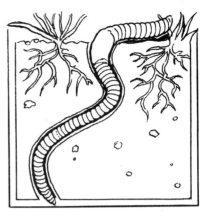

E very rain that falls through the spring and summer seasons brings multitudes of earthworms creeping to the surface of the earth to escape the rising flood waters in their underground burrows.

Why so many of them find it their manifest destiny to crawl out upon hard, paved surfaces I do not know, but wherever a street, road or sidewalk borders upon unpaved earth, there the pinkish-brown worms creep about in their numbers. And when the sun comes out, as sooner or later it does, those worms that have not been stepped upon or driven over succumb to the drying heat and lie there, drawn up in desiccated and lifeless rings.

The earthworms that inhabit my lawn and garden are no brighter than any others and, with every soaking shower, great numbers of them stretch themselves out on the flagstones at my kitchen door and on the concrete apron in front of my garage.

Because they are little wild animals, extremely valuable to the life of the soil and to all other life on earth, I try to save as many as I can. They cannot be swept off with a broom or hosed off with a stream of water without causing serious injury and death to most of them, so I airlift each little evacuee individually, tossing it gently out into the rain-wet grasses.

Not every worm. Just many. All the big nightcrawlers first because they are easiest to pick up. The lesser ones have to be teased to writhe into a lifted coil so they can be nipped up by the fingertips—squeamishness is a luxury neither gardener nor naturalist can afford—and the smallest ones can be rescued only when there is an unusual supply of time and patience.

Darwin estimated, back in 1881, that there are 50,000 earthworms per acre of land, and his estimate is still the authoritative standard. That many earthworms, Darwin said, would bring

36,000 pounds of subsoil to the surface, as worm castings, on each acre of ground every year, and in twenty years these castings would cover that same acre with three inches of new soil. Thus, he said, many ruins and many artifacts are covered over and preserved.

But more importantly, seeds, too, are covered over to grow in their season. Animal droppings and the bones of dead animals, fallen twigs and leaves and blossoms, and all kinds of organic matter are buried, to decay and add their essential elements to the soil. The very holes the earthworms bore into the earth promote the aeration of the soil, and they drain water from the surface down to the thirsty roots of plants.

Darwin wrote, in fact, that "it may be doubted if there are any other animals that have played such an important part in the history of the world as these lowly . . . creatures."

The earthworm is considered a nocturnal animal, but that is because it normally comes to the surface and wanders upon the face of the land only at night. During the day it protects its naked and vulnerable body from the drying heat of the sun by keeping to its burrow and working under ground.

It digs its burrow by literally eating its way into the earth. It pushes bits of soil with all component microscopic plant and animal life into its mouth by means of a fleshy flap on its forward, or rounded, end, and grinds it to extreme fineness in its muscular gizzard with the help of tiny pebbles picked up with the soil. The elements necessary for its own nutrition are absorbed by the earthworm's body and the remainder, enriched by a secretion of lime, is pushed out at its posterior or pointed end onto the surface of the ground.

Because the pointed end of the earthworm is often seen protruding from the hole, reaching about and feeling over the surface of the ground, it is thought by many people to be the head of the earthworm.

But an earthworm doesn't have an actual head, except that it does ingest its food at the rounded end of its body. It can travel in either direction, following either the rounded or the pointed end of its body with equal speed and with apparently equal ease.

It has no eyes, but it is sensitive to white, blue and yellow light through what are called photoreceptor cells scattered over its body. Since it is not sensitive to red light, a red filter over the lens of your flashlight will enable you to watch earthworm activities at night, whereas the usual white or yellow light would send them hurrying to cover.

An earthworm has no ears, but it has sense hairs protruding through its skin which pick up vibrations of both sound and of motion. On any warm day that I set spade into my garden soil I see nightcrawlers and all sizes of ordinary earthworms climbing out of the earth as far as four or five feet away from my spade and setting forth at their greatest speed in any direction that is away from the earthquake I am causing in their territory.

An earthworm has a sense of touch, for it definitely prefers a solid surface beneath its body, and it responds to the touch of my finger by crawling away or by squirming itself into all sorts of knots and figure eights.

It has a fairly well-developed sense of taste or of smell, or perhaps both, for when I watch earthworms at night I see them shuffling dead or dying leaves about, picking up one and dropping it for another and another, finally choosing one and pulling it down into its burrow to store for further ripening or to devour it slowly in safety and in privacy.

Often I see an earthworm stretching out from its burrow and moving its mouth around and around over the reachable earth. The first time I saw this I thought the worm was searching for favorite leaves or for other decaying vegetable material, but it may very well be gathering microscopic algae or fungi to add to its diet.

An earthworm's burrow is its place of safety, and it is reluctant to leave it to find either food or a mate. Fortunately, with its highly stretchable body, the earthworm is often able to find both within the limited radius of its body-length. Fortunately, too, it can snap back into its hole the instant danger threatens, because it has kept one end of its body securely fastened to the entrance of its burrow.

On the underside of nearly every segment of its many-segmented body the earthworm has four pairs of stiff, retractable bristles. With the aid of these bristles it is able to move forward or back upon the ground and to climb up or down the smooth walls of its narrow burrow. When a robin grabs a protruding end, the worm drives these bristles into the wall of its hole and hangs on for dear life. In the end the worm usually loses, but it does give the robin a real struggle.

When searching for food, it may anchor only a few of its extreme rear segments inside its burrow, and, when searching for a mate, only a very few of either the rear or forward segments, usually the rear, may be anchored in safety.

For an earthworm is both male and female at the same time, and it needs only to encounter any other mature earthworm of its own species in order to mate. These strange matings mostly take place at night throughout the spring, but I have often chanced upon them in the cool, moist dawnings of late April and early May.

The smooth, light band that encircles an earthworm's body somewhere near its rounded end is called the clitellum, and from it comes the egg case. Some authorities say the egg case is formed from a gelatinous mucus exuded by the clitellum, others that it is an outer layer of fine skin from the clitellum itself. Whichever it may be, it is an elastic circle that loosens from the clitellum two or three days after the earthworm has mated.

This small tube then slides forward, catches the fertilized eggs on its inner surface, and is cast off—wriggled over the forward end of the worm, I understand, though I have never seen this happen. The tube then closes up at both ends to become a tough, oval capsule about one-quarter of an inch long, yellowish-brown with a strong hint of green. And these capsules I often find, during May and June, under sticks or leaves in my garden or out anywhere in the wild.

Some earthworm specialists say that as many as one hundred tiny earthworms will hatch from the eggs in one capsule, but others disagree and say so many of the eggs are infertile that only one or two healthy but very pale new worms will emerge.

These new earthworms are so pale because they have no food inside of them and are nearly transparent. As soon as they begin to eat their way down into the soil beneath them—which they do the moment they are free of their egg cases—they also begin to develop a proper earthworm color.

It will be nearly a year, perhaps longer, before they become fully mature, reproductive earthworms; but they begin, in baby-hood, to do the plowing and the burrowing—and the burying—for which they are so highly prized by people who love the fertile earth.

Dandy Lions
on the Lawn

Ten thousand thousand dandelions gild the new grasses of lawns and fields and orchards beyond my opened windows. The winds of spring ruffle their silken florets and the light of the sun glistens on their golden heads—ten thousand thousand furry lions basking in the sun.

Authorities are agreed their name is from the French, dente de lion, meaning tooth of the lion; but authorities are not agreed, not even French authorities, on why the French have called it that. Most say it is because the leaves are so deeply and so coarsely toothed. Some say it is because the long root, when it is pulled and the skin is slipped off, is so very white and "clean as a lion's tooth." Still others say it is because of the small, sharp teeth that edge the golden rays.

I prefer the last suggestion—if they must relate to a lion's dental weaponry—but I really like plain English "dandy lion" better. Not dandy in the sense of affectation, but dandy in the sense of Jim Dandy! or just dandy! Ten thousand thousand fat little, dandy little, tawny-yellow lion cubs basking on my lawn.

Yes, I know what they do to the lawns. Their broad, thick, overlapping rosettes of lion-toothed leaves flatten the grasses, smother them out, make the world safer for dandelions.

I know that ten thousand thousand hairy gray seed heads are not very pretty, and that ten thousand bare stems sticking up when the seeds are flown are ugly on the lawn, and that the prospect of ten thousand thousand thousand new dandelion plants is appalling. . . .

But a plant which so successfully circumvents the vast array of deadly weapons man has brought into use against it deserves some measure of respect.

First come machines for mowing the lawns—small clickety-clacking hand-powered jobs and great roaring tractor-pulled behemoths, lopping off the heads of dandelions with flashing knives of steel and leaving behind green paths awash in dying gold.

But those lopped-off heads are not yet dead. Wrapped in the same green bracts that enclosed them as buds, they close in upon themselves, and, in the darkness thus provided, they proceed to develop and ripen their seeds. When the seeds are ready, the dark closets open and the wind bears the robust seeds away on the same gray parachutes that would have borne them from unlopped heads well-nourished on milky sap.

And persistent mowing of the dandelion patch results only in the disks of golden flowers rising from their green rosettes on half-inch stems, if need be, to stand below the reach of the spinning blades.

Poisoned powders are then dusted upon the earth for the dandelions to feed upon. Poisoned drinks are washed upon their leaves. Points of steel, bathed in dark poisons, are pushed into their hearts. Small, young, inexperienced dandelions succumb quickly. Stronger dandelions, their nerves badly affected, twist and writhe for weeks—and recover, many of them, unless repeated dosings are given. Even so, some plants always recover and some become more or less immune.

Then, in personal man-to-plant encounters, dandelions are dug from the soil with spades, with forks, with trowels, with knives—and if any bit of the root is left the dandelion grows again.

But angry man, amid his deified Kentucky blue and creeping fescue, may yet be grateful that dandelion determination and adaptability are so great. For mankind grows hungry and dandelions produce nutrients in quantities and qualities properly heroic to match their year after year triumphant living in the face of man's yearly resolve to wipe them out.

Only one-half cup of fresh green dandelion leaves provides, for human diets, 2.7 grams of protein, 187 mg of calcium, 66 mg of phosphorous, 397 mg of potassium, 3.1 mg of iron, 35 mg of vitamin C, and 14,000 International Units of vitamin A, 9.2 grams of carbohydrates, 1.6 grams of fiber—and only 45 calories. Spinach, alone among our cultivated vegetables, comes close in any category, and spinach has less of everything except vitamin C.

When I think of how many healthy dandelion plants I have

pulled from my garden to make space for my favorite leaf lettuce, I am chagrined. Leaf lettuce has only 1.3 grams of protein, 25 mg of phosphorous, 1.4 mg of iron, 18 mg of vitamin C, 1900 IU of vitamin A, and so on. Iceberg lettuce, cabbage, celery, all stand insipid even beside leaf lettuce, and leaf lettuce is a namby-pamby compared to dandelions.

Once, many years ago, I heard a speaker say that the best and most nutritious plants had been domesticated for garden use. I believed him because it made simple sense. Now, I wonder. If the plants man domesticated were once the best and most nutritious, did man refine them into tender insipidity? Or was tenderness the special quality man chose to begin with?

And why, when plants were being corraled, was the dandelion left out? It has been used for centuries in Asia, Europe, and America as a fresh spring potherb, a kind of wild endive; and, even more especially, it has been used as a medicinal plant. In fact, its Latin name, *Taraxacum officinale,* broadly translates to "official remedy for general disorders." And so it was. And so it still is.

Winter time, and very early spring, for most of the world's people, is now, as it has always been, beset with general debility, with illnesses, with death. Some of these troubles were and are related to the storms and the bitter weathers of the season, but most of them are caused by winter-poor diets severely lacking in vitamins and iron.

For centuries medical men and medicine men have pressed bitter juice from the roots and leaves of dandelions. They have dried the roots and leaves and powdered them. In early spring they have dug fresh roots and gathered fresh green leaves and buds and golden blossoms. They have concocted teas and tonics and tablets and have given dosages of these to patients suffering from scurvy and from anemia and from various other ills, and the rapidity of their cures has seemed miraculous. Iron. And vitamin C. And vitamin A.

And all of it for free! From dandelions that have no need to be weeded or hoed or mulched or coddled. They grow luxuriantly in neglected fence corners. They do well enough on lawns where they are walked upon daily and run over weekly by portable guillotines. And they flourish like the green bay tree in fertile vegetable gardens.

Dandelion leaves and buds are delicious in fruit or vegetable salads or as salads by themselves. They are excellent as an addition to practically any soup I can think of, and they make a good "mess of greens" when cooked like spinach and dressed as you please. Roast the roots for coffee, or boil them like parsnips. And use the full blown flowers for wine.

Ancient lore says the leaves should not be eaten after the plants flower, but that is only because of a toughening of their texture. Shred them finer for salads and cook them longer for greens and you can use them the whole summer through. Besides, if you keep using the tiny buds the plant never has a chance to flower.

Modern lore says that no part of any plant should be used if it or the area in which it grows has been dosed with a weed killer in any recent year. I can suggest no way of getting around this sage advice. I do not know how long the root retains the poison nor how much of the poison gets into the leaves from roots and from soil poisoned a year or two ago. Perhaps only a little trace of poison is concentrated in the seeds, so you can gather the seed and raise your own Jim Dandy Lions in uncontaminated soil.

The Pageant in the Garden

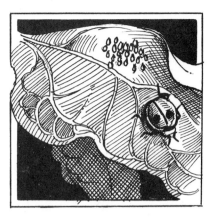

The great thrumming procession of insect multiplication is in full swing these late spring days, an unflagging pageant of action and color, of intricate designs, of plots and counterplots. I am abroad, and on the *qui vive,* with a magnifying lens held firmly in hand— and I scarcely need stray beyond my own garden.

On the undersides of tender lettuce leaves I am happy to find small clusters of sharply-pointed oval eggs, tangerine-colored, and standing on their tips. Ladybug eggs. I've seen only a half-dozen aphids in the garden this morning, but they were winged ones, soon to give birth to dozens and hundreds of sticky green daughters. They will, inadvertently and finally, provide a livelihood for the tiny blue-and-tangerine dragons now sharpening their claws inside those tangerine eggs.

Other small clusters of sharply-pointed oval eggs, these yellow in color, and set up on their tips on the undersides of the young bean leaves, do not enthrall me. They are the eggs of the Mexican bean beetle, a vegetarian member of the ladybug tribe, whose mandibles are equipped with several teeth—the better to eat the fine-haired leaves.

A double row of small white eggs, banded with black, on one of the larger radish leaves takes me by surprise. Under the lens they look like small white barrels with their hoops painted black. I recognize them as the eggs of the harlequin bug, although I have never seen them before. But then, I have seldom seen a harlequin bug, either. Harlequin bugs, the full-grown ones, are either a shining black or a deep, deep blue, so lavishly marked with red that they are often called calico bugs. They are an importation from Mexico, and they thrive on all the mustards.

The golden-eyed lacewing is a delicate, pale-green insect with transparent wings too large for its body and a pair of eyes as strikingly golden as the name suggests. Her eggs, airily resting on filmy pedestals, are scattered everywhere. Several, in a single line, climb the heavy stem of a pea vine; a few are clustered on one pea pod; a cabbage leaf boasts a dozen or so; there are some on the hollyhocks, many on the roses.

I do not see a lacewing in action this morning, but on other days I have seen her egg-laying procedure. First she lowers the tip of her abdomen to the surface of the leaf or of the stem and she ejects a drop of sticky fluid. Then she lifts her abdomen half an inch or so, pulling the fluid up in a fine thread behind her. The thread hardens in the air as it rises, and she immediately ejects an egg, pinhead size, oblong, and white, onto the top of the thread.

The egg is filmed with the sticky fluid, the tip of the thread hardens to the egg, and the egg is firmly attached to its hair-thin pedestal. Thus each egg is placed safely out of range of the voracious young "aphis-lions" who will hatch from her eggs and who will climb down from, but rarely up, the surrounding pedestals. This extraordinary care does not, however, prevent the mother lacewing herself from absentmindedly eating a few of her own new-laid eggs.

Cabbage butterflies do not eat cabbage leaves, although their young devour them; so the white butterfly I see has no reason for fluttering and hovering about my half-dozen cabbages except to place her far-too-many yellow eggs there. To my unaided eye her eggs are simply raised dots of yellow on the cabbages, but through the lens they appear beautifully formed, like a flask, and they are marked with twelve or so vertical ribs, though not all of them extend to the uneven, scalloped top of the flask.

A black swallowtail butterfly, the most common swallow-tail in our area, is placing her smooth, round, yellow eggs, one by one, here and there on the fine-leaved dill. I watch her closely. There are carrots and parsley in the garden and Queen Anne's lace on its borders, all of them legitimate food for black swallowtail larvae, but she remains faithful to the dill. Apparently, and from my own observations only, the egg-laying female chooses one species of the umbelliferae and deposits her eggs on that species alone.

I know that the larvae, the caterpillars, will eat only the species on which their eggs were laid and on which they make their first meal. They will starve to death if the supply of that leaf is depleted, even though they are surrounded by other species which should be food plants for them. (This information, also, is only from my own observation.)

The eggs of the tiger swallowtail are smooth and green and not-quite-round, and they are placed by the gracefully soaring female on the leaves of such trees as the birch, poplar, ash or wild cherry, or on the shad bush. The food plants of her larvae are said to be more varied than those for any other butterfly in this country, but whether each female places her eggs only on the ash or only on the birch, I do not know.

A few monarch butterflies are filtering back to our area now, some of them with orange-and-black wings so tattered and worn that they look as though they have come the whole distance from Mexico. Perhaps they have. But the consensus of opinion seems to be, these days, that they may even be *second* generation from the hibernators and may have migrated only a few hundred northerly miles to lay their eggs on our freshly burgeoning milkweeds.

Monarch eggs are easy to find, with or without the egg-laying female to guide you. A single, upstanding, white dot on the broad surface of a milkweed leaf is plainly visible to the naked eye. But, again, you need a hand lens to see that the egg is shaped like a sugarloaf and has twenty-one or twenty-three slender ridges radiating from top to base. The spaces between the ridges are marked with parallel lines, and, tying the whole design together, a tiny rosette is delicately etched into the center of the top.

The eggs of the mourning cloak butterfly are oval and yellow, and marked, vertically, with eight or nine ribs. The mourning cloak female, who has overwintered as an adult, will lay her eggs, in May, in one large and sometimes disorderly cluster encircling a twig of a willow, an elm or a poplar tree. But the females that develop from this brood will, later in the summer, place their eggs, on end, in neat rows, on the underside of a leaf on the same kind of tree. From this second brood, the adults will overwinter and, the following May, lay their eggs in untidy clusters around a willow twig.

A relative of the mourning cloak, commonly known as the violet tip, produces perhaps the most beautiful eggs in butterflydom. Each egg surface is simply divided by ten ribs that extend from the center-top to the base; but they so refract the light that the eggs gleam like clusters of tiny jewels—like green-lighted jewels before the caterpillars hatch and like tiny diamonds when the shells are empty. These glittering gems are placed at staggered intervals on an elm tree, a hop vine or a nettle, so that the finished arrangement is a very loose cluster of eggs.

Far from clustering her eggs, the red-spotted purple butterfly lays each one of her minute green eggs singly, on the upper side of the extreme tip of a cherry tree, apple tree, shadbush or rose.

And so the procession goes on in endless ranks, with butterflies and damsel flies and dragonflies; with water bugs, with fireflies, with water skippers; with ants and hornets and beetles and bugs ad infinitum. Every mated female is laying a quantitude of eggs which, in their placement, in their numbers, and even in their colors, are laid with an eye to confounding their predators and to continuing and increasing the multitudes of their species.

Wherever you may be you are standing in the midst of a pageant of action and color, of elegant designs, of intricate plots and counterplots. Don't miss the performance.

The Massacre of the Giants

Someone is killing off all the great old snapping turtles that lived in Heron pond. They are killing them off, I'm sure, with the best of good will toward the pond and the owners of the pond. Their benevolent intent is only to protect the fish that inhabit its waters and to insure good fishing for the numerous fishermen who drop their lines from its banks.

They, whoever they may be, are not taking these turtles for food. If they were, the deed could be understood and even found excusable, for these turtles are huge and meaty and a real temptation to anyone who has ever banqueted on a delectable turtle steak or tasted a savory snapper stew. But this is wanton, wasteful slaughter. The turtles are simply being killed and left on the banks to rot.

The first one that I discovered was shot cleanly through the head as it rested at the surface of the water breathing in a fresh supply of oxygen. Possibly because its lungs were filled with air, this turtle did not sink to the bottom when it died but floated on the top of the water for days under the flattened dome of its impressive carapace.

Other snappers by the half-dozens have been dragged ashore by one means or another and have there been brutally beaten with clubs until they died. The dark carapaces of the two largest of these that I found measured fourteen and one-half and fifteen inches, respectively, and when the one floating out in the center of the pond finally drifted into shore I measured its carapace at sixteen and one-half inches in diameter.

The carapace (upper shell) of a turtle is not measured by following its curves but by a straight line from front to back, so that the official measurement makes the shell an inch or so smaller than the eye can see it really is. Thus, the carapace of the average adult snapping turtle measures, officially, only eight to twelve

inches in length, which seems awfully small to me. But the all-time record, so far as I have read, is a carapace only eighteen and one-half inches long.

Now, snapping turtles do not grow to even the eight-to-twelve-inch dimensions unless they have a satisfactory supply of food, and the great hulking size of these old fellows indicates that their years in this pond have been years of total abundance.

Even if their entire diet were fresh-caught game fish, which it isn't, their heavy, fleshy bodies, protruding in fatty rolls from under the edges of their smoothed-down carapaces, would be ample evidence that there are plenty of fish in this pond for everybody.

But fresh fish make up only about one-third of a snapping turtle's diet. They eat a surprising amount of vegetable matter, mostly the roots, stems and leaves of aquatic plants; they eat small water birds, reptiles and mammals; and they eat an unbelievable amount of carrion.

And what kind of carrion do the snappers find to eat at this pond? They feast lavishly upon the bodies of the undersized bluegills that disappointed fishermen toss back into the waters on every weekend and special holiday, on every lovely morning and evening, while fishing weather lasts.

These multitudinous little bluegills are tossed back so they may continue to grow into pan-sized fries, but the mouths of many are badly torn when the hook is removed, or their innards are crushed by ungentle handling, and so they die, and float repulsively on the waters that lap the sand bars and the grasses on the edges of the pond. It is upon these bodies that the snapping turtles feast, and, in so doing, keep the pond waters clean for all its inhabitants and its edges from becoming a reeking sump.

Resident snapping turtles fill a definite niche in the life of any pond or stream, and in this pond they are sorely needed, not just for scavenging, but for consuming some of the overabundant small bluegills. There are huge bass which eat bluegills, and there were, in years past, huge snapping turtles, but still the bluegill population, I am told by fishermen, is overwhelming.

Truly, it is hard to love a snapping turtle. They are not pretty in either face or figure and their belligerence and pugnacity, when out on the land, makes them anything but appealing. Their reputation for "eating all the game fish" is highly exaggerated, but their reputation for ferocity is not. Whether they can "snap a broomstick in two" I do not know, but that they can snap off a child's finger I have no doubt.

Whether or not they can bite through the fingerbone of a full-grown man, again I do not know; but their necks are thick and muscular, their heads are large, their toothless mouths gape widely open, and their powerful jaws are rimmed with bone and as sharp as razor blades. They can, without question, deliver a grievous wound.

As captives, or out on land, they are adversaries to be seriously considered, but as free creatures in their watery habitat they are usually docile. If you step on one in the water, it will not attack. It will just draw in its head and sink unobtrusively down into the mud—not because it is afraid, I suspect, but because it is so placidly secure in its own element.

Out on land, the snapping turtle is extremely vulnerable. It has the smallest plastron (undershell) of any turtle in America—only half a dozen scutes (large scales or plates)—so that its underparts and its appendages are almost completely unprotected. It is probably to compensate for this weakness in its structural design that the snapping turtle lunges and snaps at anything that stirs in its vicinity when it is ashore.

But, like the skunk before it sprays, the snapping turtle goes through a warning maneuver before it attacks—if, of course, there is time. The turtle rises high on its hind legs, flexes its joints, elevates its long, saw-toothed tail, and lifts the rear of its carapace; it opens its mouth and lets its lower jaw swing agape; then it lunges and strikes, repeatedly, and with lightning speed. Even a baby snapper, scarcely larger than a twenty-five-cent piece, will lunge and snap as soon as it is dry from the egg.

From babyhood through adulthood there is no danger of mistaking a snapping turtle for any other turtle in our area. Their heads are large, their plastrons are small, their upper shells are rough, and their long tails, as long as their carapace or longer, are saw-toothed like a dragon's tail. Razor-beaked, tiger-clawed, turtle-shelled, and dragon-tailed, they have a savage disposition and a lightning-fast attack, and they are not to be fooled with, at any size or any age.

Snapping turtles may mate from spring until fall. They mate

on land, usually quite close to home waters, and I find them, in spring and early summer, on the earth dam at the pond not more than half a dozen feet from the water's edge. The female, however, may wander quite some distance inland, a thousand feet or more, before she digs her nesting cavity in the earth and deposits her twenty to twenty-five white, leathery-shelled eggs in the dark pocket.

Usually she chooses the loose earth of a bank or the cultivated soil of a grain field or garden in which to dig her four-to-six-inch cavity with her unaccustomed and practically unadapted hind feet; but, once, I came upon a snapping turtle female in a drought-ridden month of August attempting to dig her nesting pocket in the hard-baked clay-and-gravel of a roadside ditch.

Now, I fully believe that there is no living creature on this earth that is a more unaware, mindless automaton than the female of any species of turtle when she is occupied with her nesting and egg-laying activities, and this female snapping turtle was no exception; but I left her strictly alone. If she had been any other turtle species of my acquaintance I'd have picked her up and set her on the more favorable earth of a bank not five feet away and she'd probably have gone on digging quite unconscious of having been moved. But I don't mess with snapping turtles.

Given the regulation ninety-day incubation period for snapper eggs; those, like hers, laid late in August would normally hatch sometime in cold and inhospitable November, while eggs laid in September and October would hatch in the even less hospitable months of December and January.

But nature takes care of these little problems. The increasing coldness of the nest cavity apparently inhibits the development of the little turtle embryos so that they are not ready for hatching until the warm and welcoming days of the following spring.

In fact, it is said that, in some years, the baby turtles hatching in late August and September may find the earth so hard-baked and dry above them that they cannot dig their way out of their nests. In that case they simply estivate and hibernate until the warm and welcoming days of the next year's spring before they make their appearance on the face of the earth and, in some unknown and most remarkable manner, find their tiny ways to the murky waters of the old home pond.

Perhaps there may have been a nest or two of these overwintering baby snappers out in the hilly cornfields around this pond. And perhaps they found their ways home, so that these waters will not for long be bereft of their snapping-turtle population.

But, even so, it will be many a year before the inch-long babies with their inch-long tails attain the weight and the size of the irascible old giants so recently dispatched. And until these babies grow up (or some overlooked half-grown ones do) there will be something primitive and wild and savage—and necessary —missing from the deceptively tranquil waters of the pasture pond.

The Ecological Mosquito

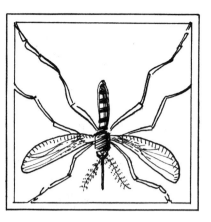

The soft, warm closings of our summer days bring starlight and cricket song, dewdrops and fireflies, nighthawks and owls and whip-poor-wills, elusive, eddying fragrances, and unwelcome mosquitoes in uncountable numbers.

The instant the high rim of the sun drops behind the hills, they rise on gossamer wings from shadowy lodgings in bushes and vines and down-drooping grasses. Their fragile bodies float lightly on the twilit air and they follow the changing scent trails that lead to the springs of their warm nourishment.

However it may feel to the human skin, not every mosquito is out for blood. And, of those who are, great numbers choose or stumble upon the blood of dogs or cats or foxes or cattle or rabbits. A blood-seeking mosquito is an opportunist, and it takes its meal from the first available creature it finds on its evening flight.

But, whatever the species or the gender of the host, the mosquito taking the blood meal is always a female.

The bushy-antennaed male mosquito confines his drinking to fruit juices and the nectar of flowers. Female mosquitoes, too, until they are mated, are content with nectar for their food. But, once mated, they thirst for blood, are implacable in their quest. Not to preserve their own lives but to insure the generations.

For their own health, female mosquitoes can live as heartily upon nectar as upon blood. They can even generate a few viable eggs. But in order to produce the regular complement of one-hundred to four-hundred eggs for each laying, every female must have a meal of animal blood, and it must be a full meal. It must be a meal adequate to turn her abdomen red and distend it to its extreme capacity.

If she is disturbed as she eats, she may have to bite several times through the night until she is replete. And she can be disturbed easily for her whole body is so exquisitely sensitive that the very tensing of your skin as you *think* about swatting her feels, to her, like an earthquake, and she is gone in a fraction of a flash.

Once satisfactorily filled, though, she retires from active life —or, at least, she retires from eating—and she devotes the next few days, perhaps a full week, to digesting the blood and replacing it with eggs. It is as though the blood and the eggs produced with its aid are in exact ratio for her abdomen remains excessively distended throughout the generative process.

When her eggs have ripened she finds some quiet water in which to lay them. This is the second of the undisputed facts about mosquitoes, world wide (the first is that the mosquito that bites you is a female): All mosquitoes lay their eggs in water, or, some few, in a low spot which will soon be filled with water.

Some kinds of mosquitoes, and there are around two-hundred species in North America, lay their eggs while doing a bobbing dance, up and down, just a few inches above the surface of the water. A female of the common, domestic, *Culex* variety either sits directly on the water skin or finds a bit of floating debris for support, and she drops her eggs behind her, one by one, all in the same spot.

But she keeps pushing them about, to one side or the other, with her hind pair of legs so that her several hundred eggs are formed into a raft afloat at the top of the water. The eggs are covered with a sticky substance that hardens as it dries and the eggs of the raft are glued firmly to one another.

These rafts of eggs turn out in all manner of shapes and sizes, and, so far as I know, there is nothing species-specific about the form. The raft configuration, I think, rests with the lap of the water and the chance kick of the leg.

And now that she has a batch of eggs safely deposited, this female mosquito, depending upon her species, upon her individual strength and upon the season of the year, may simply die. Or she may search out another host, gorge herself with another meal of blood, and produce another batch of eggs. Or, if winter approaches, she may have that second meal and go into hibernation.

The male mosquito drifts through summer nights of dalliance, sipping nectar from small flowers, possibly piercing the thin skin of a plant stem now and then for a healthy meal of sap, and quietly filling his own minute ecological niche. For others, beside his own species, are dependent upon him. There are, for one thing, several species of very tiny wildflowers, flowers like the forget-me-nots and the dainty Quaker ladies, so tiny that the bees do not bother with them, which may be entirely dependent upon mosquitoes for their pollination.

This vegetarian male mosquito dallying among the blossoms probably feels no hunger for animal blood, but if he did he could not indulge his whim. His undeveloped mouth parts are simply not equal to the job of cutting through animal skin.

The female comes equipped with two fine needles that vibrate like dentists' drills as they cut through epidermis, into the dermis, and then into a handy capillary. Her saliva, flowing down her cutting tools, chemically and instantly thins the blood it touches, so that, as she drinks, it does not clog the extremely fine hollow of the sucking tube she inserts from within the narrow circumference of her two vibrating needles.

All of this almost microscopic, fine-precision equipment she carries within a protective sheath which is, itself, a tool. The sheath actually penetrates the skin first, just slightly, and then, on a pair of knee-type joints, it partly doubles up on itself and allows the needles to complete the job. And this sheath, pressing its lobed tips against the skin, steadies the rest of the apparatus so that the whole business of obtaining the blood meal is quickly and efficiently done.

To pin-point the differences in the mouth parts of the male and the female mosquito requires the use of a very strong lens, but the differences in their antennae can be observed with the naked eye. The antennae of the female are plain or only slightly branched, while the antennae of the male are as feathery as the antennae of a male moth.

But the feathery antennae of the moth are organs of smell and the feathery antennae of the mosquito are organs of hearing. The male moth tracks down the female of his species by means of her perfume, but the male mosquito pursues the siren quality of the music produced by his prospective mate.

Whether or not they hear sounds other than the one sound of their destiny, I do not know, but they do hear the song of the female mosquito. That faint, whining song that chills the skin and curdles the blood of the human in his bed sends the male mosquito into a frenzy of delight.

Surprisingly, that sound, that song, is not made by the swiftly vibrating (from 278 to 307 times per second for the common *Culex* variety) wings of the female, but by the rippling of thin scale-like projections that lie across her breathing pores.

Mosquitoes are such an annoyance and such a danger to man in their hordes, and even in their dozens, that their total eradication, he thinks, would not leave him sorrowful.

But man puts up great castles of martin houses and rejoices in the numbers of mosquitoes his resident colony of graceful dark birds devour, and he fails to note the other side of the coin, that the martins are dependent on a constant and goodly supply of mosquitoes for their livelihood.

Thousands upon thousands of birds and bats and dragonflies truly depend on the mosquito hordes as a prime source of their sustenance. They swoop through the air gobbling adult mosquitoes as innocently and with as little malice as a herd of cattle pastures upon meadow grasses or as a swarming of mosquitoes feeds upon animal blood.

While we, peering shortsightedly into what makes the world go 'round, look with innocent delight upon the graceful turnings of the swallow's flight, we view with searing malice the gossamer floatings of the mosquito mobs. But this is one more of those many things we cannot have both ways. If we clear the air of pesky insects, we clear the air of swooping birds.

Came the Deluge

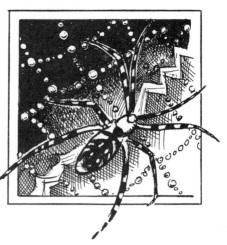

For several years the rains had been plenteous, with never a flooding, never a too-dry spell, and the bustling, prolific, aromatic marsh stayed wet and puddly through the fertile seasons. Stayed wet and stayed in bloom—a continuous helter-skelter flower garden, from spring's lovely, almost-hidden blossoms to the riotous gold-and-purple glory of late summer and fall.

The creek waters, those years, flowed twelve to eighteen inches deep between its broad banks, rippling over shallows, plunging through narrows, running silent and solemn through deep, dark pools, providing space and conditions for many kinds of lives.

Under the creek waters, in small breaks and caves in the banks, thriving colonies of crayfish, from tiny transparent babes to green, five-inch adults, enlivened every turning and every shelving in the creek.

Frogs were there in abundance, in and out of the waters. Bull frogs, green frogs, leopard frogs and pickerel frogs. Tadpoles of all their kinds in the spring, and tadpoles of the bull frogs from one year's end to the other.

And there were fish. Wild fish. Native fish. Flickering schools of minikins, all eyes and tails, and six- to ten-inch hiders-under-rocks and darters-into-weedy-corners. Brook trout and dace, for two, I have been told, but I do not recognize any fish by name.

Congregations of water striders skated on the surface of every quiet back water and drifted on the currents of every silent pool. Swarms of hard-shelled whirligigs gyrated in nonstop circles atop the water, and backswimmers, strange creatures, lying on their backs, rowed themselves hither and yon.

Above the waters, many-hued damsel flies drifted and colorful dragonflies zoomed, while, under the waters, their eggs and their larvae slowly developed toward the state of becoming the next summer's generation of flyers-in-the-air and layers-of-indis-pensable-eggs.

Every August and September twelve to fifteen giant fishing spiders hung their silken packets of eggs in the greenery overhang-ing the waters of the creek. Each mother constructed a balloon-shaped nursery, then sat on guard below it until the baby spiders hatched and wandered away and left the nursery empty.

In these same months thousands of orb-weaving spiders hung their large, delicate webs on the green growth all about the marsh. Three species were predominant among them: the brown Araneus of the rounded abdomens, the striking black-and yellow Argiopes, and their smaller relatives, the pale, banded Argiopes. All of these spiders, in August or September, placed their masses of pearly eggs on vegetation near their webs, wrapped them warmly in silken blankets, and left them to become the spider population of the following summer.

And all of this was only a part of the continuity and the variety of lives among the unfurred and the unfeathered in the creek and marsh community.

Then came a summer when practically no rain fell on the small valley. The water in the creek dwindled until it flowed only one or two inches deep in the very lowest levels of the creek bed. As their living quarters grew smaller, the inhabitants of the creek crowded closely and more closely together, under the water, in the water, and on top of it. Wandering frogs and salamanders and water, snakes crept back to the creek from the drying marsh and the congestion increased as conditions worsened.

In the silted bottom of the now waterless creek bed, on either side of the narrowed stream, sunflowers and Joe Pye weed, jewel weed, iron weed, asters and goldenrod sprang up, grew thick and lush, and thrust their heads high above the banks.

In mid-August, for the first time in years, a tractor roared into the marsh and mowed down all its flower-garden richness and color, all its reeds and its grasses, all its gossamer spider webs and all its population of spiders. Only the tall growth in the creek bed was left standing, and it became, now, a long, narrow flower border threading through the dying blossoms and the sharp stub-ble of the marsh.

Hundreds of spiders who had survived the massacre in the

marsh found their way to the tall vegetation in the creek bed, and within the week, great wheels of webbed lace were hanging at every conceivable angle and in every available space in the flower garden of the creek bed.

Now all the swarming, lowly life of the marsh was concentrated between the two banks of the meandering creek. The stage was set for disaster, and disaster walked in on cue.

In the first week of September a violent thunderstorm raged through the valley, drowning it with inches of pounding rain. Mud-filled, pebble-loaded waters churned out the waterways, sluiced through the ditches, boiled down the creek channel, tearing stones from their moorings, undercutting and overflowing banks, scouring out silt and plants and animal life, swirling it, beating it, carrying it away. The creek and the marsh were denuded.

But, by the very next summer, the marsh had returned to its full flower-garden glory, and only someone deeply familiar with its every face could know that three of its rarest species were missing: The great blue lobelia, and both the Maryland and the Virginia meadow-beauties. All three of these lovely flowers grew close beside the creek in a saucer-shaped hollow that was cut out and washed away by the grinding waters of the September flood.

Four years have now gone by since that devastating storm, but since the morning before the deluge, I have not found one giant fishing spider or seen one of their nursery webs along that stretch of creek. Nor have I found a single wild fish in the stream. Not in four years!

The orb-weaving spiders, the big, brown Araneus and the pale, banded Argiope, are, only now, slowly making a comeback. But the handsome black-and-yellow Argiopes are scarcely to be found.

A few crayfish, those few that were hidden in deep caverns not broken down and swept out by the rampaging waters, survived the storm. But in the four years that have passed, they have not increased their numbers. Where once a dozen could be seen outside any one of their underwater pueblos, I now cannot find that many in the entire length of the creek. And none that I see appear to be of reproductive size.

The whirligig beetles are gone completely, and only a small

rowing of water boatmen remain. The groups of graceful water striders are missing from the stream, and in the air above it only an occasional dark dragonfly zooms and scarcely a damsel fly drifts by.

For the first two years following the storm, I walked along the creek banks morning after morning and evening after evening without causing a single frog to leap into the water. The third summer there were increasing numbers of small, naked-looking froglets leaping from the banks. This spring, the fourth spring, masses of eggs floated on and in the waters, swarms of tadpoles followed them, and this summer the whole place has been jumping and twanging with froggy forms and voices.

But the bullfrog story is a sad one. All those beautiful frogs with voices like river steamers are missing from the banks of the stream and from the pasture pond itself. Every new season I keep hoping that some great, green-faced adventurer will come hopping upstream looking for unclaimed territory; but if any have, or do, they are keeping their presence a secret from me.

Such was, and is, the depletion of the richness and the variety from among the very lowly in this creek and this well-watered marsh in only one hour of storm.

I tell this tale because one of its elements differs from the usual iliad of disappearing species: Except for the mowing of the marsh which sent the orb-weaving spiders to the vegetation in the creek, it was natural forces, not the hand of man, that brought about the concentration and the destruction of those hosts of little lives.

The Real
Daddy Longlegs

The British have always called them "harvestmen," and the name is fast catching on in this country because they are mostly seen at harvest time and because "daddy longlegs" sounds like a fanciful title dreamed up by a child and scarcely to be used in dignity by an adult.

We, as children, always called them granddaddy longlegs, but I've never found that designation in print or heard, in my adult years, anyone else use the term. It may have been only a local identification, or, even more likely, a label bestowed, and exclusively used, by my own large and quite unregimented family.

Some people call the long-legged, and somewhat clumsy, crane fly a daddy longlegs, and others call any long-legged spider the same thing, so perhaps the real daddy longlegs does need a name that is all its own.

The real daddy longlegs belongs to the Arachnid class along with spiders and mites and scorpions and ticks. Its small (two-tenths to three-tenths of an inch) gray or brown body has all its parts broadly joined so that it appears to be just one small round or oval blob cantilevered in the center of eight, long, exceedingly thin and very fragile legs.

It is neither a rare nor a vanishing breed, yet not so common a sight as not to be noticed when one does appear. In my childhood, the cry "Here's a granddaddy longlegs!" always brought several of us on the run to watch, and sometimes, daringly, to touch that little body lightly bouncing and swaying among all its fantastic legs.

Some authorities say those legs, so easily detached, will grow again if lost. Others say they are not replaceable. I can only say that in a lifetime of daddy-longlegs-watching, I have seen many hobbling about with one, two, or even three legs missing. I have never seen one with a short, regrowing leg. Spiders, yes. Daddy longlegs, no.

The second pair of legs is much longer than the other pairs. When the daddy longlegs moves about, this pair of legs "walks" too, but it keeps in rapid, dancing motion over the surface upon which the creature travels.

The tips of this pair of legs are extremely sensitive. They are the feelers that test the texture and the safety of the path. Dark patches on the tips of this pair of legs can be seen with the unaided eye. These dark patches are the organs for smelling and for tasting. This pair of legs also does the hearing. At each joint there are rows of very fine hairs which pick up vibrations from the air and, possibly, the ground. A daddy longlegs does not hear sound in the sense that an animal equipped with ear drums hears sounds, but it "hears" what it needs for its kind of living.

Quite obviously, a daddy longlegs that loses one of this second pair of legs is seriously hampered, and if it loses both of them it is lost. Every sense, except the sense of sight, is located in the organs of these two legs. On the top of its body, located between this second pair of legs, the daddy longlegs wears a small black dot. This dot is actually a knob, a small turret from which, on either side, a shiny black eye looks out in a lifelong, unblinking, unchanging, sidewise stare.

A daddy longlegs breathes through small holes in its sides, called spiracles. And it eats through a small slit of a mouth in the very front of its body. The daddy longlegs first locates food by touching, smelling and tasting it with the special organs in its second pair of legs. Then it tests this finding with a smaller pair of feelers (modified legs) located at the front of its body. If it really is food it has found, another pair of very short modified legs, located on either side of its mouth, picks up the food with pincer tips, crushes it, and stuffs it into the mouth.

Daddy longlegs are omniverous. They eat small living insects which they capture or dead ones which they find, fresh or rotting fruits and other vegetable matter, and they are attracted to droplets or puddles of water, for they seem to be thirsty all the time.

They cannot walk on top of water but they can stand on it for short periods of time—as long as their clawed feet do not pierce the water skin. When that happens their wet legs become tangled and the daddy longlegs quickly drown.

Pounding rain can kill these fragile creatures, too, as can the hailstones that sometimes accompany a summer storm, but they seem to have no living enemies—perhaps because of the odor they carry.

Behind the first pair of legs there are two odor glands which look like eyes, but which give off a smelly liquid when a daddy longlegs is annoyed or afraid. To a human being this odor is very faint. I have never smelled it, but it is said to smell like walnuts. For some reason this odorous substance is offensive to cats, birds, toads and large, predaceous insects. Not only is it offensive, it apparently makes some of these predators feel weak or nauseated.

And so daddy longlegs are said to have no enemies. But I have discovered that their lives are not completely pest-free. In the late summer and throughout the autumn, of some years, almost every one I find is carrying what looks like a tiny, blood-red, liquid droplet on one or more of its legs. My hand lens reveals that this droplet is actually an eight-legged mite, a pest, a drainer-away of daddy longlegs energies.

The time of the mites, late summer and early fall, is also courting time for daddy longlegs.

Daddy Longleg courtships are calm, straightforward affairs. They are not preceded, so far as I can discover, by any ritual dances. They are not complicated by either the bizarre mechanics or the wild rapacity of spider matings, nor are they followed by attempted murder or achieved cannibalism. Daddy longlegs simply mate.

Soon after mating, the female daddy longlegs lays twenty or thirty pale green eggs in a rotting log, under moss, behind tree bark. She has, surprisingly, a long protrusible ovipositor, and she places her eggs, one at a time, in her chosen hiding places.

The hatchlings—tinier than a pin head, but looking exactly like adult daddy longlegs—do not break from their eggshells until the warmth of spring sunshine triggers their need to get out and stretch.

They grow very quickly, I am told, and they shed their first skins before they have spent an hour in the open world. They hang onto the tree bark or the moss or a grass stem while the skin splits down the back. Then they struggle out, resplendent in a new, loose skin that looks exactly like the old.

They don't shed again until they have spent ten days or so

eating bits of decaying animal and vegetable matter and, if they succeed in capturing any, extremely tiny insects, even smaller than themselves.

They continue to shed their skins approximately every ten days until, after about nine sheddings, they reach adulthood.

Presumably, this tenth skin is their last and they do not have to go through the discomfort and interruption of shedding their skin again.

The baby daddy longlegs, because they are so small and so very shy, are seldom seen; but, sometime in June, half-grown ones wobble across the porch or over the raspberry leaves with increasing self-confidence and little need to scurry.

By early August they are full-grown. They sway and bounce sedately in the midst of all their two-kneed legs, and they take on a solemn and venerable air that makes the title granddaddy longlegs seem most appropriate.

In late summer or early fall, then, they seek out mates and they place their resultant eggs in sheltered places against the winter's cold, but they are not yet finished with their own living. While the weather remains warm, they eat and drink and bounce about in their normally separate ways, but, as the evenings grow chill, they begin to congregate, particularly, it seems to me, in the corners of my windows, where, probably, warmth seeps through from the house. They pile up out there, one on top of another, until several groups of ten or a dozen are nested together looking like small and hairy (because of all the legs) inverted baskets.

Before the days turn really cold, these long-legged creatures wander off to creep under leaves and stones or into hollow logs; their life systems slow down almost to the vanishing point—and some of them live through the winter.

It is not very likely, I'm afraid, that any of them survive through a second winter to become, in actuality, granddaddy longlegs.

A Mantis' Last Measure

The air was warm that November day, the sky was clear, and the bright sun shining against the south-facing wall of my home made a little island of summer there among the frost-touched roses and the fading chrysanthemums.

A ladybird beetle bustled to the tip of a rose leaf, then turned and bustled back again. A paper wasp queen, not yet settled for the winter, wandered without purpose up and down a window pane. A wolf spider dashed across sunlit, bare earth to hide in the shadow of a gray stone. Three honeybees hummed in the late purple asters; a belated monarch butterfly paused there, too, for a fill-up of aster nectar before drifting on toward the south. And a female praying mantis, even more belated than the butterfly, wandered slowly among the tightly grown twigs and branches of a small-leaved holly shrub at the corner of the house.

The autumn's frosts and freezes had taken their toll of the mantis. One wing cover drooped, the other was torn. All her green coloring had turned to bronze, and most of her bronze had changed to russet-red. She walked heavily, wearily, dragging her grossly distended abdomen across the branches, deliberating, with slow head turnings and enormous eyes grown dull, upon every aspect of every twig and every branching stem.

I don't know how long she had been there before I saw her, but she looked so exhausted, so driven by this last measure nature demanded of her, that I was certain the laying of her eggs was minutely imminent.

But three hours and twenty minutes went by before she stopped her nervous wandering and stood rocking forward and back on a spot she had examined at least twenty times before. She was clinging head downward to the stem of a strong main branch, about ten inches down from the top of the shrub, on the inside toward the house wall. Mantis instinct, I thought, has chosen a proper site—or perhaps she can't wait another minute.

But she could wait another forty minutes, rocking there on her thin, angled legs. Even the saw-toothed front legs, usually drawn up beneath her chin, were stretched out before her, reaching down the stem, balancing and supporting her bedraggled body.

Darkness fell. The glow from the lamplight within the house fell softly upon her. The mantis slowly pressed her body backward, straightening the angle of her legs, not loosening her claws, not moving from the site, just pushing her abdomen upward as far as it would reach.

A gluey fluid flowed from the tip of her abdomen onto the smooth-barked stem. Tiny, hair-like appendages on either side of her abdomen's tip began to whirl like egg beaters whipping bubbles of air into the whitening liquid, and the abdomen itself rotated slowly, slowly, slowly.

She whipped the sticky fluid into a thin, flattened, upstanding round of heavy froth that glistened in the lamplight and then, almost imperceptibly, she eased her body forward just a fraction of an inch. More fluid flowed from her abdomen. The tiny beaters whirled, and, as the bubbling foam took shape, the mantis eased down the stem, bending her legs, not moving her feet.

More fluid, more beating, and, with the third or fourth thin level, the egg case began to take shape with ornamental ridges down its sides and a band of closed gateways to its inner passages at center front. Slowly, slowly, slowly the egg case grew.

The hour was nearing midnight—the temperature dropped only to the mid-sixties—when the mantis pulled her flattened abdomen from the last foam addition to her inch-and-a-quarter egg case, and stood with her wasted body balanced in a normal position above the thin angles of her legs.

She brushed one folded front foot across each eye and then walked with slow, deliberate steps straight down the stem and stepped off on to the brown-leafed earth. She stood there for one moment. She swayed forward. Swayed back. Swayed forward again, and her six legs folded so acutely at her sides that her long body rested down between her feet, flat-out on the all-supporting earth.

Next morning, somewhat after sunrise, she lifted her body, dropped back to earth again, and two hours later she was dead.

Her egg case of hardening froth, still white and conspicuous, was moored firmly to the branch of the holly shrub. I eyed it with intense curiosity. How were her eggs arranged in that case?

There are mantis-watchers who say, in print, that the mantis lays an egg or two every minute throughout the process of forming her foam-insulated case. But surely that would mean the eggs would be whipped haphazardly through the case and the new hatchlings would be held prisoner in the hardened foam when they came out of their eggs in the spring, if, indeed, they could get out of the embedded eggs at all.

Other watchers say the mantis lays her two-hundred to three-hundred eggs all in one cluster in a central cavity of the case, and others report that she lays a small cluster of eggs in each level of the foam case as she builds it. Either of these methods, it seems to me, could be true. And it might be that not all mantises build their cases or lay their eggs in precisely the same manner.

I could have settled the matter for myself, probably, by slicing through that, or some other, egg case, but I have never had the temerity to destroy an egg case just to satisfy my curiosity. It seems to me that the eggs of the mantis tribe have enough dangers to contend with—the vagaries of the winter weather, the ripping bills of sparrows and chickadees and titmice, and the tearing teeth of little rodents—without having additional perils from me.

I was tempted, though, in this instance, for my research here would probably destroy nothing. How much life could so weary, so belated a mother mantis possibly have deposited in that store of eggs?

Fortunately, I did not open the now darkening case, for nearly seven months later, on a sunny morning in May, I noticed that the gateways in the narrow band down the front of the dry egg case were spread slightly open and that the dark, shiny eyes of tiny new mantises were peering out into the light.

When next I looked, half a hundred new-hatching creatures, soft and damp and yellowish, were dangling down the holly stem all hooked together by silken threads in a chain-like garland of mantises. They came spilling out of the gateways, willy-nilly, whether they wanted to come or no, the weight of the living chain pulling each newcomer over the threshold and out into the sun-shine of the great spring world.

Somehow, one by one, and probably by accident, they broke the silken threads that bound them to one another, they cleared themselves from the tangle of knees and legs, and covered the branches and twigs of the holly shrub in a creeping dispersal of almost-transparent, miniature mantises, all awkwardly trying out their spindly, unmanageable legs.

And just as suddenly as the mantises had hatched, nature's population controls set to work to reduce their numbers. Ants of all sizes stalked everywhere among the multitudes of soft new mantises, attacking with merciless jaws of iron the tender bodies of the confused and helpless infants. Sparrows and chickadees gathered in great excitement and feasted noisily on the tasty delicacies clustered so in one spot. There was Maytime mayhem outside my window.

But judicious nature again was taking a hand. The outer cov-ering of the baby mantises swiftly hardened, making them safe from the ant-hill hordes. And with the hardening of their exo-skeletons came green-brown coloring and an awareness of the world they'd inherited.

They slipped away from the ebullient sparrows and chick-adees into a haven of greens and browns—green grass, brown stems, green leaves, brown earth—where they, almost invisible, preyed in their turn on soft-bodied creatures more helpless than they.

Just a Little Green Bird

The tiny green bird, scarcely larger than a humming bird, sat at just about my shoulder level among the green needles of a spruce tree high on the cold, cloud-shrouded top of Mt. Mitchell in North Carolina.

Its eyes were large and dark and shining, and circled by such prominent white rings that the bird looked wild-eyed and scared to death. But it sat there on the down-mountain side of the trail I was following, within three feet of my own staring eyes, and quietly and calmly returned my gaze without ruffling a feather or blinking a tiny green eyelid.

I didn't have a field guide with me because, one, my load limit for really enjoying a walk of any distance—half a mile or twenty miles—is approximately three ounces; and, two, I was such a novice at bird watching in those days that I thought I could just see them and look them up in the book when I got back to camp, or back to the car, or whatever.

Finding the birds was the big job, I thought. Identification was a simple matter of matching up a picture with what I remembered seeing in the woods and fields and marshes.

And so it had proved—up until now. Birds were so diversified in color and pattern that identification was no problem whatsoever. Cardinals, goldfinches, chickadees, titmice, scarlet tanagers, flickers. Bird watching was a breeze.

And so I came down from Mt. Mitchell's cold, gray summit and, that evening, while the gray rain pelted down from the gray heavens, I curled up in a motel room, warm and dry and cozy, and leafed through my field guide in pursuit of the little green bird.

I found pictures of several greenish birds among the flycatchers, kinglets, vireos, and warblers. All of them had white rings around their eyes as well as two white lines on their wings—I couldn't remember whether my little green bird had white lines on

its wings or not—but all of them looked far more gray than green and I knew that my little bird was *green.*

I leafed on through the book until, almost at the very end I found one little green bird, a female painted bunting and I heaved a sigh of relief and of faith-and-diligence-finally-rewarded. True, the eye ring in the picture looked a little less prominent than on the real bird of the mountain top, and the eye seemed a trifle small, and I rather remembered a dainty black bill instead of the cone-shaped yellow-green bill the painted bunting lady was wearing, but, still and all, it was a fairly good likeness of my little green bird in the spruce tree. It was a little bird and it was green.

But when I turned to the descriptive text I discovered that painted buntings are inhabitants of southern towns and southern thickets and that they travel no further north than southeastern North Carolina. And that settled that. I could make allowances for discrepancies in the picture well enough, giving a little here, allowing a little there, but the inhospitable top of Mt. Mitchell was no southern town and its climate must be at least as northerly as northern New England. My little green bird was not a female painted bunting.

But what kind of bird could it be? Was it possible that I—a bird watcher as green as the little green bird—had discovered, on my first real expedition, a new species unknown to the learned ornithologists of eastern North America? It seemed quite likely.

I didn't know, then, that new bird watchers make it a common practice to discover totally new species of birds, sometimes at the rate of one a day.

Nor did I know, then, that Roger Tory Peterson is the ultimate depictor of birds, that when he paints a bird with a cone-shaped bill the bird has a cone-shaped bill. And the same goes for eye-sizes, and eye-rings, and eye-stripes, for wing-bars and tail-shapes and leg-colors, for crests and crowns and anything else that makes up the bird. Individual birds within a species may vary a little in one detail or another, mostly in intensity or area of coloration, and there are, of course, birds that are albino or melanistic in great or little degree. But, except in the matter of precise color, Mr. Peterson means what he paints.

(And so do other field-guide artists. What the printing process does to their colors is not their fault. I have one bird guide now which is textually excellent, but all of its birds, even its sparrows, are green. What a search I would have had through this book for my little green bird.)

The years went by and I did learn a few things about one bird or another. I learned about habitats and nesting habits. I learned about bills and tails and wing-bars and eye-rings. I learned to take occasional color variations in stride. I learned to know some birds so well that a fleeting glimpse or a bit of song identifies them without thought on the matter. But I did not, for a great many years, learn the identity of my little green bird.

One cold December day I stood at my bathroom window watching two little gray-green kinglets skitter about in the olive trees just outside, not bothering a bit about the mockingbird's treasured red olive berries but daintily skimming microscopic insect eggs and larvae from the leafless stems, flitting lightly from one branchlet to another, fluttering like stocky butterflies over twig ends, and sometimes hanging upside down from a swaying sprig.

After some little time I roused from my absorption in their wiry call-notes and in their flittering feeding on things I could not see, to a sudden awareness that the feather-coats on these withy kinglets were not gray-green, they were *green*. As green as my little green bird in the spruce tree on Mt. Mitchell's foggy peak. Their eyes were large and dark and shining, and their prominent white eye-rings made them look wild-eyed and scared to death. A kinglet! My little green bird had been a ruby-crowned kinglet! I was as elated as though I had found an emerald.

I saw that little green bird in October. It must have been an early migrant from summer nesting in the cold spruce belt of Canada, keeping to the mountain tops in its southward journey, along with others of its kind, I'm sure, to prolong its sojourn in a northerly type of weather.

Fortunately for us, not all of the kinglets follow the mountains south, and, even more fortunately, hundreds of the little fellows elect to spend their winters in Maryland. Some may winter as far north as New England, others spread out through all our southern states, and some journey across the Gulf of Mexico to the hospitality of Guatamala.

Few of us ever get to see the ruby crown on the head of the

male ruby-crowned kinglet because he keeps it well hidden under the fine, tiny feathers that cover it.

Birds are named from dead specimens, and it may be that those who christened this kinglet did not know that the ruby crown would not be visible as a field mark. In death the covering feathers fall aside and the brilliant crown is a prominent feature —a fact I have learned sadly and at first hand over the years, from the several birds that have done themselves in by accidentally striking against the windows of my house.

Males are said to fly their ruby flags when they are battling one another for nesting territory up in the northland among the Canadian spruces: but this, of course, I have never seen.

I did once, though, see the little male of that very green pair in my olive trees flashing his bright crown in and out, while he hopped about twittering his wiry winter song notes as he prettily lifted and fluttered his wings before the tiny female who stood quietly watching him.

Although kinglets often lift and flutter their wings while they trip about among the twigs, I took this for a courtship dance because the time was early March, he confined his hopping to a very small area just in front of the female, and, shortly afterward, he opened his thin little black bill and poured forth the merriest, bubblingest, and loudest song that I ever heard coming from such a teeny, tiny throat.

Only a few days later all the kinglets had vanished from the shrubby environs of lawns, orchards, pastures and woods edges.

With mates, perhaps, already chosen they were off to the Canadian spruce belt to build their pendant, globe-shaped nests; to pile eight or nine creamy, brown-speckled eggs into each narrow, mossy bottom; to brood, to hatch, to feed, to fledge their little ones; and to wing back safely, we hope, to spend another winter in the southern U.S.A.

Feather Trails

A blue jay feather, richly blue and tipped with white, lay at the edge of the woods atop the fluff of snow that had fallen during the night. A tail feather, stiff and straight, its blue marked in runic black, its blade-shaped surface neat and trim. How came it here, loose, and unowned, and lying by the wayside? January is not a proper molting time for birds, but how else explain so pristine a feather lying on smooth, untrampled snow?

From the stiff midrib of this feather hundreds of fibers (barbs) extend on either side, giving it the look of a slender, bright blue leaf. These barbs are so fine, so closely set together, that they seem, of themselves, to make up the web or the vane of the feather. But with my hand lens (and the feather held up to the light) I can see the forests of even finer branches—there may be a million of them on this one feather—set closely along the edges of the barbs.

These fine branches, called barbules, closely overlap the barbules of the adjoining barbs and are interlocked with them in a straightforward system of hooklets and barbicels so tiny my lens does not reveal them. But I know they are there because I have seen them under the lens of a microscope.

I gently pull two barbs apart. I can feel the light resistance as the hooklets unhook among the forests of overlapping barbules. Now I try to rehook them. As I said, it's a simple straightforward system. Each barbule is hooked to, and hooked by, the barbules on either side. And they need only to be pushed together. The birds do it all the time. Without hands. Without fingers. With only two stiff mandibles.

Just recently I read, in a book for children, that the barbs of a rumpled feather can "easily be slipped back into place" by laying the feather flat and drawing it slowly through closed fingers.

Well, I've been trying that system for more years than I'd like to state and I have, occasionally and accidentally, succeeded in rehooking a few barbs, but never smoothly, never completely, never tightly enough to shed a raindrop. However, I'm not noted for my manual dexterity. Perhaps feathers can easily be hooked together again by hands with the proper touch.

But feathers don't rehook themselves. The barbs, once pulled apart, stay apart, until or unless the bill of the bird preens them into smoothness on its own body—or a most dexterous hand restores a fallen feather.

But this feather needed no restoring. Not a barb on this feather was unhooked from its fellow. Neither tooth nor claw nor talon had ripped this feather from the tail of an escaping jay. Surely no violence brought it down. Perhaps, like furry creatures, a bird now and then sheds some of its covering out of season. . . .

I walked farther into the woods. The snow lay cupped in curled brown leaves and it tufted lightly over a mound of pale, downy feathers pulled from the breast of a surprised barred owl.

I know the barred owl was surprised, but it should not have been. It had, for some reason I cannot fathom, taken shelter, at ground level, in a thin-shelled, hollow stump open from the side but not from the top. The owl must have sheltered there more than just once, for I found several of its regurgitated pellets of mouse-hair and mouse-bones lying just outside the narrow entrance to the hollow stump, and a double-handful of the pellets lying inside on the leaf-deep floor of the shelter itself.

I don't know how well the owl was eating, how many pellets it would be likely to disgorge each day, or each night, so I can't gauge, by their number, how often it sheltered in the hollow stump before, last night, a fox discovered it and pulled it forth.

The owl must surely have been asleep, deeply asleep. And the fox must have grabbed it by the breast and pulled it through the narrow opening while it was still too unaware to open its wide wings, else there would have been blood, and damaged wing feathers, and a broken stump. But there were none of these. Only a great mouthful of brown-and-white feathers, soft as down, on top of the leaves and under the snow.

The leaves were scattered, here, and ruffled, and out of their flattened-down-for-the-winter alignment, but only within a radius of half a dozen feet from the entrance to the hollow stump. No other feathers littered the ground. And no fur, either.

I think that, once clear of the splintery entrance, the owl came awake and flailed the fox with the power of its wings, beat it about its head, and began at the same time to pull away in an aerial escape. The hold the fox had on the owl's breast must have been mostly on its feathers, and the feathers gave way in the tug-of-war between the downward pull of strong fox teeth and the upward pull of great owl wings.

The owl must have been a young one, inexperienced and unacquainted with furry enemies, or it would never have sheltered in so precarious a spot. Perhaps the fox was young, too, and inexperienced, or it would have grasped its prey more firmly at the start. At any rate, last night, before the snow fell, both an owl and a fox learned a lesson in survival. The lesson cost the owl its breast feathers, and it cost the fox its dinner.

We left the woods, then, my dog and I, crossed the fallow field of meadow mouse fame, and entered our neighbor's marsh. We splashed across the swollen creek and climbed the long hill toward the sassafras copse. Halfway up the hill we came upon the scattered black feathers of what once had been a crow.

All the feathers were there—down feathers, contour feathers, flight feathers, tail feathers. Not in a heap, but wildly scattered. Not neat and trim, but ravaged, ragged, pulled and wet. Tracks of fox and tracks of wings marked the furrowed and winnowed cover of snow.

I don't know why the crow was on the ground in this spot nor how the fox crept upon it across this open, grassy hillside. But this I do know: One crow lost a battle and one fox ate crow for dinner.

Now the sun broke through the clouds, low over the hills in the east, and the flat whiteness of the snow became ridged and broken and blue with shadows.

Among the gray rocks on the edge of a ravine the two-two tracks of a white-footed mouse, like the tracks of a dainty, lilliputian rabbit, were sharply etched in the lowlit snow.

We followed the traces in their adventurous wandering down into the ravine, up the bank, up and over the white brow of the hill. And there they ended. On either side of the last dainty footprints the marks of spread wing feathers, of hawk or of owl, were firmly pressed into the tell-tale snow.

Untimely Ladybug

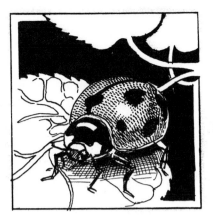

On a snowy February afternoon a little round insect with a high-domed back came prancing out of the woodwork in my study like a song-and-dance man out of the wings, and balanced on the edge of the windowsill. It teetered there on its thin black legs, preened its short antennae, and surveyed however much of my household world it could see with its finely faceted eyes.

I looked upon it with amusement and with reproach for I never know what to do with these ladybugs who overwinter inside my house and then break off their hibernation before the winter is gone. If I put them out of doors they will probably freeze to death before they find a safe cranny, and if I leave them to their own resources, indoors, they will probably end up in the dust bag of my vacuum cleaner.

Why they come back to life so inopportunely, when the weather outside is still deadly and the temperature in my house is no higher than it has been on any other day, I do not understand. Some physical factor must trigger their little alarms and set them scurrying about on summer-type errands, but, though I note their appearance every winter, I can find no particular catalyst, no common thread, except that they do appear.

Sometimes it is the little two-spotted ladybug (*Adalia bipunctata*) who spends the winter in my company, but this one is a nine-spotted ladybug (*Coccinello novemnotata*) and she represents, I think, the most common species in my territory.

Both of these little beetles have tangerine wing covers so bright and hard they look like baked enamel, and both are dotted with black. The two-spotted, of course, has only one black dot, prominently worn, on each cover, while the nine-spotted wears a polka-dotted design with three prominent black dots on the rising curves of each wing cover, one fainter or smaller dot low on each side, and another dot on the very top of its high-curved back, one-half of the dot on each wing cover.

But the arrangement of these markings is not invariable. I look closely at this ladybug and see that the two smaller dots are faint and smudged and placed so low on its wing covers that they are almost off the edges. The center dot, shared by both wing covers, is also faint and its margins are blurred, but the little insect remains a bona fide nine-spotted ladybug.

Its body is more round than oval, its back is high-domed, its head practically hidden under a hard, curved scale, called a sclerite, which extends forward from the base of its wing covers.

There on the windowsill, the little ladybug raised its hard, bright wing covers and exposed, beneath, a little round body in unrelieved black. It shook out a pair of tiny wings in filmy chiffon of the dullest black, rose on the tips of its toes (tarsal claws), then dropped back, folded its wings, settled its wingcovers over them, and strolled, afoot, across the windowsill and into a bookcase, where it vanished into a barren canyon between two volumes of the Brittanica.

Quite safe from vacuum cleaner nozzles and hastily wielded dust cloths and protected from desiccation by its hard wing covers, its danger, if it remains there, now lies in being crushed by books carelessly handled. If only I will now remember my responsibilities, the beetle may make it through to spring to lay at least one clutch of pointed, tangerine-colored eggs.

The tangerine coloring is a signature that runs through every phase in the life of the nine-spotted ladybug. First the narrow eggs, standing on end in a close-packed mass of twenty to fifty, make a small splash of tangerine on the underside of a leaf— lettuce leaf, cherry leaf, rose leaf, almost always on a plant that does, or that may, harbor sticky herds of aphids.

Once, long ago, I chanced upon a ladybug laying the last egg in a cluster on the underside of a lettuce leaf and thus learned for certain whose eggs these were. I kept a close watch on them for I didn't know what a ladybug larva looked like, and, after several days, fearful that I might miss their hatching, I brought the leaf into the house and kept it fresh by standing it in a small glass of water.

Even so, I failed to see the hatchlings leave their eggs. I came into the kitchen early one summer morning to find the egg shells empty and the leaf, the glass, the countertop all a-crawl with tiny, spined, many-segmented dragons, pointed at each end, wide in the middle, and irregularly colored in pale blue and tangerine. I slipped a sheet of paper under the wanderers as gently as I could, hoping I was not injuring or crippling their tender new bodies, and I took them out and shook them into the lettuce bed.

Tiny as they were, and new to the world, they promptly recognized aphids of any color as their proper food, and they hunted them out in swift, erratic and continuous wanderings over the lettuce leaves. They were voracious little savages, those growing pastel dragons, and they did not hesitate to nip my skin when they found themselves on my hand or arm and momentarily unsupplied with aphids.

They grew wider and longer and tougher-skinned, and when they were perhaps a quarter-inch long, they disappeared. I knew they had gone off somewhere to pupate and to turn into adult ladybugs, but, at that time, I didn't know what procedure they followed and I didn't know where to look for them.

It was not until the following summer that, quite by chance, at ten o'clock in the morning, I discovered a lilac-and-tangerine ladybug larva hanging head downward in the center of a cherry leaf. A hardened droplet of clear glue-like substance at its tail held it fast while its six legs clawed frantically and continuously at the leaf surface.

Just before noon the larva succeeded in pulling itself free. It moved somewhat to the right and a little farther down on the leaf, exuded another drop of clear glue, settled for a few moments until the glue was fairly hardened, then began to claw frantically as before.

At three o'clock in the afternoon the larva again freed itself from the restraining glue and moved down to the point of the leaf where it exuded still another droplet of glue and settled down, this time to stay. Its six legs clawed slowly at the leaf-surface but the action seemed more a reflex motion than a frantic struggle. At intervals, then, for several hours, its whole body heaved forward and back, up and down. The dragon shape began to disappear, as though both pointed ends of its body were pushing themselves into the wider middle.

The next morning a little, humped, beetle-looking shell sat there, silent and still, its tail pressed, shriveled and flat, beneath it, its lilac-and-tangerine color lightened to pale beige.

Twenty-four hours later the faded larval skin had split over the front and was slipping back over a hardened body in tangerine-touched-with-black to hang in crinkled folds around its cemented tail.

For the next three days the beetle-marked pupa shell hung on the tip of the cherry leaf, apparently lifeless, shielding the remarkable transformation going on within it.

On the morning of the seventh day I found the pupal shell split and empty, and, beside it, an adult ladybug little more than an eighth of an inch long. Its body was round and strongly convex, and its black body-color showed dimly through its thin and not-yet-hardened wing covers.

Less than an hour later I watched it move out, in brilliant tangerine spotted with nine black polka dots, to feast on a herd of black aphids pasturing on the leaf stem. It was far less voracious as a ladybug adult than as a ladybug dragon. It consumed fewer aphids, more quietly, with less dashing about, but it still remained their formidable enemy.

For ladybugs and all their close relatives, so far as I know, are indefatigable eaters of their species-specific foods throughout both their larval states and their adulthoods—as witness this nine-spotted ladybug, the Mexican bean beetle, and the squash beetle, to name a simple few.

Merry, Merry Bobolinks

ifteen years ago, on a blue and silver morning in the middle of May, Kon-Tiki—my first Great Pyrenees puppy—and I ran through the pasture fields in light-hearted adventure. With careless feet we went, scattering dandelion seeds from the fat round fluffs that grayed the meadows where, for all these spring weeks, the bright golden disks of dandelion blossoms had glistened in the sun.

We had slowed our pace to climb the steep, briar-tangled hillside beyond the pond, heading for the higher levels of the hilly pasture land, when suddenly the air above us was filled with light, tinkling, silvery music, the glittering notes speeding faster and faster and faster, as though ten thousand fairy pianos were being played at a gallop up on that enchanted hilltop.

As we neared the top, I peered wonderingly over the rim and saw, among the low forest of dandelion grayheads that topped the grasses there, a numberless throng of black-bottomed birds, their backs patched and striped with white and with yellow, avidly feeding upon the freshly ripened seeds while they bubbled and tinkled their merry, merry notes.

Bobolinks! Of course! But how many years since I had heard them! And never, never had I heard more than three or four together—and here were ten thousand at a glance! I held Tiki to my side and sat down on a rock below the rim just to listen to their light-hearted fairy tinkling.

There is something about the summery merriment of just one bobolink's song that drives poets, and even staid writers of prose, to extravagant lengths. Thoreau wrote that ". . . he (the bobolink) touched his harp with . . . liquid melody and . . . the notes fell like bubbles from the strings." Among the many poets who tried, Wilson Flagg wrote a long, rollicking poem that captures the spirit of the bobolink's rhapsody, and even the melancholy William Cullen Bryant responded to the lively mirth of the bobolink with

a surprisingly brisk (for him) "Robert of Lincoln." But George Gladden probably said it all when he wrote, simply, that the bobolink "sounds like an hysterical music box."

And maybe he does. His song begins on a low key and goes joyously skylarking upward and it does have a slight but definite metallic quality. Maybe the tone *is* more like a music box than like a fairy piano. . . . But, then, I heard a bobolink long before I ever heard a music box, and when did I hear a fairy piano?

So I mused as I sat on my ringside rock, quieting a restive Tiki and listening to this great glee-club convention of migrating male bobolinks who had paused here to feast on dandelion seeds, before moving on to their more northerly nesting grounds.

There were so many of them, and every one singing solo at the same time, that I could not determine whether they were singing a travel ditty, or a grace-while-eating, or if they were, perhaps, polishing up their courtship songs; for they were certainly all wearing their courtship plumage.

Every feather of their underparts, from chin to breast to tip of tail was deep, deep black. The same deep black colored their bills, their faces, their eyes, their heads, their tails. But wide across the backs of the heads and the backs of the necks, reaching to the shoulders, lay a wide cap of soft bright yellow.

Their backs, from shoulders to waist, were dressed in longitudinal stripes of black and yellow, and a fine line of yellow edged each of the stiff black feathers of their wings. Their scapulars (on the upper edge of each wing) were purest white, and so were their upper-tail coverts. They are the only American songbirds who, during the nesting season, wear dark below and light above. They look exactly, as Roger Tory Peterson says, as though they have their dress suits on backward.

But, however striking their appearance, it is their frolicsome singing that I love, so I closed my eyes and luxuriated in the never-before multitudinous merriment about me.

But, just as one cannot remain forever on the mountaintop, neither can one rest for very long on a hilltop bubbling with bobolinks, especially when accompanied by a Great Pyrenees impatient for her walk. So Tiki and I proceeded with our morning ramble and returned to our prosaic world where I dutifully marked the date in my journal—May 19.

On the following morning there were perhaps a hundred male

bobolinks among the nearly depleted dandelion seed-heads on the pasture hill, and on the morning following that only a desultory two dozen or so. Then I saw and heard no more bobolinks, not even the dull-brown-striped females who surely flocked through a few days later.

A year went by and the pasture fields were again grey with dandelion seed-heads. Running there with Tiki, I gave no thought to bobolinks. Then a flock of them flowed up from the masses of fluffy spheres ahead of us, and their tinkling tunes chimed from the little black throats of a thousand birds, lined up on every tree and bush and fence wire in the vicinity. It was May 19! The swallows of Capistrano have nothing on the bobolinks of Dawson's pasture, I thought, except, of course, that the bobolinks are only passing through. . . .

The next year I was looking and listening for bobolinks. A few of them were in the pasture on May 16, on May 17, on May 18; but on May 19 a great host of gaily melodious, industriously feeding bobolinks landed there and tradition was safe for another season.

The next spring and the next spring and the next spring on the nineteenth of May the pasture fields were crowded with dandelion grayheads and with migrating male bobolinks in classical dress and jubilant voice.

But the following year—and now I walked with Kela, for Tiki had died suddenly, and very young—the main crop of dandelions went hurriedly to seed in April, and the great May flight of male bobolinks did not alight where there was no feast spread for them, but flew on to the north to catch up with their ripening food supply.

And so, through the years, the pattern evolved: If a great harvest of dandelion seeds was spread for their feasting, the main flights of the male bobolinks fed and rested here on and about (but always on) the nineteenth day of May.

But if the crop was early, or if it failed—which sometimes, unbelievably, does happen—the male bobolinks passed us by. I reasoned that the great flights of the females, which I never saw, probably did not stop here at all because there was never, at their later flying-over, any great crop of seeds ready for their feeding.

Dandelion-seed-time has now become a watched-for calendar event in my life, and, when it falls properly, I wait in happy anticipation of hearing yet once more the merriment of the tinkling-voiced bobolinks, and I live in barely suppressed anxiety and dread lest they should change their itinerary or time their arrival for another date.

But they never do. Though other changes come—I walk, now, with Kuonny—May 19 is still the day on which the great wave of feasting and frolicking bobolinks greets me on the pasture hills.

This year, once again, I'm watching the dandelions and waiting in anxiety, and in wonder, too, because last year, on May 19, 1979, the bobolinks threw a surprise party that took my breath away.

On that bright morning, the intensity and the volume of bobolink music in the pasture was so great that we heard it before we left our own woods. And when we reached the meadowlands we found not only the hilly fields but the hollows, the marshes, the trees, the bushes, the hedges, the lanes and the fence wires practically submerged in bobolinks, noisy and merry and feasting and waiting and flying and perching wherever they could find a body-sized space.

Thousands upon thousands of black-bottomed, fancy-dressed male bobolinks, and thousands upon thousands of sparrow-drab, brown-striped female bobolinks; all of them in one great spread flock; all of them together; all of them tinkling metallic calls and musical tunes; and all of them glistening dully in the brilliance of the rising sun.

Something drastic must have happened to the schedule of female flights that they should thus overtake the main passage of their males. And something drastic must have happened to the supposed instinctual psychology of their chastely separated flight times that they should fly thus merrily northward together.

On May 19, 1980. . . .

Note to my readers: The main crop of dandelion seeds ripened ten days early in 1980, and while I saw small flocks of merrily tinkling male bobolinks on May tenth, eleventh and twelfth, I saw none at all on the nineteenth.

However, this essay appeared in the Baltimore *Sunday Sun* on May 4, 1980, and every one of my readers must have watched the

subsequent ripening of the dandelion seeds and looked for the
arrival of the male bobolinks, because I was practically inundated
with calls and notes from delighted people who had never seen or
heard a bobolink before. But whether they were new to bobolinks
or old hands at their migrations, people kept asking me, the whole
month long, "Did you see them?" "Did you hear them?" Everybody was sharing bobolinks!

In the Darkness of Evening

The sun had dropped behind the hills, and the sky, above a high scattering of cloud-mist, was the palest of pale evening blues.

Alone, I walked along the margins of the drought-lowered creek attempting to see into its shallow depths, but its surface was only an opaque mirroring of the pale sky and I saw the same thing below me as above. Now and again small frogs leaped from the bank ahead of my loitering feet and shattered that reflected sky, but the ripples closed so swiftly over them that all I actually saw was the shadow of a motion.

Abandoning the creek, I climbed the briar-grown rise of the earth dam, crept down its curving pond-side almost to water level, and seated myself, silent and out of sight, among June's high-reaching weeds and grasses.

But the pond world knew I was there. As quiet and as accustomed as my approach had been, my alien human presence created a disturbance in the wild world of the pond margins and I had to wait while the broken peace restored itself and the life I had interrupted could resume where it left off.

Slowly then, I simmered down, quieted my mind, became absorbed in watching the darkness settle upon the waters.

Darkness did not come, as I had thought it would, in a slow and steady lessening of light as the sun moved westward across our nation—(or rather, as the earth whirled eastward.) Darkness came in waves, in layers, in the sudden dropping of netted curtains one after the other.

Between fallings of curtains a last winging of barn swallows flocked low across the pond. As their lines swept twittering before me, a dozen swallows dipped their bills for a drink, cutting sparkling slices of silver spray into the darkling water.

Then a green frog struggled upward through a patch of matted green algae just off from shore. He pushed through its bubbly gelatinous surface and sat there in remotest silence with a thimble of its green piled on the top of his head and strings of it dangling wetly down his back—the algae-crowned king of a floating green island.

When, at last, he moved, I thought he was making an exploratory hopping tour of his kingdom. But, all at once, he dropped, waggle-legged, through a bit of openwork in its green edging and came up several feet away, still trailing a tangle of green strings, like mermaid hair, from his froggy head. He swam quickly to the edge of the earth dam and sat under the curved-down grasses at the water's edge, where I could no longer see him.

I could hear him, though. He, too, seemed fascinated by the layered manner in which night-time was arriving, and, as each new wave of darkness settled down upon the water, he saluted it with one, short, explosive, congratulatory "Twunk!"

Each succeeding darkening lay on the water for an appreciable length of time before the next degree drifted down, and the last one had turned the water into a black mirror in which the great oaks near the inlet, the willows, the grasses, the long slope of the eastern hill and the rough face of rock on its broken side were all clearly and deeply reflected.

The next thickening of the dark dropped with such abruptness and such considerable depth that it brought a loud, surprised, rubber-band "Twunk! Twunk! Twunk!" from the green frog below me and a sudden series of responding "Twunks!" from a host of green frogs all around the pond.

One lone fish leaped from the water and flashed silvery scales in light I had not thought was there.

Now, from directly below me, from the water at the very edge of the dam, the expanding silvery V of a muskrat's bubbly wake cut toward the distant inlet. I knew a furry brown head cleaved the water at the apex of that V, but I could not see it and the water seemed parted by a magic knife. The muskrat swam straight away for more than half the length of the long pond, then rolled and dived and disappeared; but the heaving waters reported that he moved restlessly in its depths.

Soon he, or another muskrat, surfaced somewhere toward the upper end of the pond and the apex of its shining wedge swept down the pond toward the dam. I softly breathed away my excitement, quieted my scampering brain, and watched the muskrat end his swim with a rising through the surface rubble at the edge of the dam and a dripping clamber onto its solid earth, to settle down among its grasses not twenty feet from where I sat.

In the almost palpable dusk that washed about us, the muskrat appeared only an indistinct furry form on a dim green bank, but I could plainly hear the crisp biting, the tearing, the shredding as he dined on his green pasturage.

He sat in that one spot and pivoted about his own haunches, confining his grazing to an extremely small radius. He was either completely unaware or completely accepting of my presence, for he ate without hurry, paused now and again to rub his face or to look about him as a rabbit does, but he showed not the slightest sign of uneasiness or of agitation. After a long while, I have no idea how long, he rubbed his face rather thoroughly, scratched his left side, then waddled down the curving dam, slipped into the water, and was gone.

By this time the gnats and mosquitoes were pasturing freely upon me, and I moved out of the grasses, over to the west bank, to sit upon the dry, exposed, breeze-touched wooden pier.

This rustling and creeping moving-about silenced the green frogs and the one bullfrog of the evening who had just begun his river-boat whistle from the other end of the pond, changed the minds of any muskrats who might have considered coming out upon the banks, and ended my opportunities for observing unguessed happenings. But I have never been much convinced I should feed the mosquitoes.

So now I inspected the water beneath the pier. Thousands of shining little fish—the fish all heads, and the heads all big round eyes and gulping mouths—were feeding at the surface with tiny, sucking, wet-mouth noises. Suddenly, without warning but with a tinkling of finny splashes, they turned invisible tails and fled off through the black water.

I lifted my head. Against the sky, I saw the pair of resident mallards come winging across the now-vanished dam and fly up the inky pond far too high for a landing on this passing, but, just this side of the inlet stream, they abruptly halted their flight in mid-air, tipped themselves up, and dropped vertically, in reverse, down toward the dark, invisible water.

Their bills pointing to the sky, their wings beating downward, braking against the air, their feet reaching, reaching toward the water, they dropped through the dusk-filled night.

Tails flaring, wings drumming in shorter, blunter strokes, webbed feet feeling, feeling for the water. Their toes dipped into its wetness, their bodies dropped forward, and, for just one instant, I saw their horizontal forms afloat by the sand bars before the darkness swallowed them.

Now water and earth and sky were dark. There were summer constellations of stars in the sky but they were dim and small and far away and not one twinkle of their light reached down to light the pond or the pasture.

I stumbled from the pier to the top of the dam. Below me the marsh stretched away in solid darkness. A week from now the marsh and the pond margins would be awash in firefly light, but tonight only a single flickering here and there marked the presence of a firefly that had arrived before its time.

I have walked out on many nights, before and since, on nights that should have been darker, on moonless nights and cloud-shrouded nights, and I have found my paths and my ways easily and plainly visible, sometimes most startlingly so. But this night was the darkest night I have ever experienced. It was not a dark night of the soul. It was not a darkness psychologically induced. I was not frightened. Not afraid. Not even slightly so. I knew where I was and I loved where I was. It was simply a dark, dark night, and why it should have been so thickly black I do not know, but I certainly could not see the ground whereon I walked.

I pressed toward home. The creek was a meandering blackness on my right and the marsh a lumpy murkiness to my left. I walked on the paler strip of dry land between them, and I crept without mishap beneath two invisible electric fences, though how I knew where they were I have never been able to determine.

A narrow path, barely six inches wide and innocent of any landmarks even if I could have seen them, led diagonally across the old sheep pasture. With shoulder-high orchard grasses and goldenrod leaning across it, this path, unless I strike it properly at its beginning, is difficult to find even on a sunny morning. On this night it was impossible. I could not discern the grasses, let alone the six-inch division of a path running through them. I could not even guess where it ran.

But my feet found the path, and, as long as I kept my eyes from peering, my mind from seeking, they followed it unerringly, until my hands touched the solid planks of the wooden gate, and I climbed over.

Again, my feet followed the path my eyes could not see across a clearing in the eastern end of the woods, but when I plunged into the really dark tunnel of the woods-road, I was lost. Not lost in the physical sense—I knew where I stood, knew where the road lay, but I could not find my way along it.

Now this woods road is five or six feet wide, but it is strewn with fallen branches, pocked with protruding stones; and several small trees, six or eight inches in diameter, have fallen across it. It is not easy to negotiate in the dark, and turning off my mind wouldn't work here. My feet just could not find the way.

And I wondered about that as I felt my way along, arms extended, feet testing every step with exceeding care.

The path between the creek and the marsh follows the meanderings of the creek, wandering and curving to conform to the local contours of the earth down the marshy way. It was originally tramped down by cattle, or probably before the cattle came, by wild animals and by Indians following the animal trails.

The path across the sheep pasture is not so ancient, but it was first made by wild animals—by deer, by foxes, by raccoons— seeking, and finding, a way from woods to marsh to creek to pond that avoided the impassable obstructions set up by man. This path also wanders and bends and dips and climbs and follows the sometimes almost imperceptible changes in the local contours of the earth.

Both of these paths belong.

But the woods road does not belong. It dips and it climbs but it does not wander. It marks the shortest route between the county road and the tenant house that once stood in the clearing at the back of the woods. It is a road of expedience that, even without its blockages, would require eyes to follow.

I spent a long time stumbling down the short, rough length of the woods road that night—listening to the skitterings of wood mice among the old leaves, breathing the rich rising odor of evening-damp woods earth, being startled by the cry of a barred owl from somewhere just south of my path.

But at last I found the piles of stone near the end of the road and crept out from under the low-hanging dogwoods onto newly mowed grass where the forgotten light from my study window made a broad beacon across the open lawn.

This light I followed quite easily to my wide front door and the welcoming walls of home.

Dragonflies at Noonday

In the shadows of thin-leaved willows and of tall-grown Joe Pye weeds, damsel flies with dull black wings and bodies of metalic green or blue fluttered in short and dancing flights, or rested with their wings clasped tightly above their backs in rearward pointing angles. But out in the heat of the noonday sun, powerful dragonflies zoomed among the reeds or flashed over the earth dam that held back the pond.

They hunted for mates and they hunted for food, and they found them both. They mated by flying tandem, with the male in the lead, his abdominal claspers holding the head of the female in a caveman grip and her abdomen curving forward in a loop beneath them. They fed by folding their legs into baskets and straining the air for all kinds of insects. They dined on the wing without missing a beat. They dashed at top speed in a dozen directions and erratically changed course without signal or sign.

Only when these vigorous dragons paused, with wings outspread, to rest upon a twig or to hover over a puddle beside my path, could I any more than glimpse their fantastic patterning.

Most of the dragonfly bodies were black, and all of the dragonfly wings were gauze—clear, transparent and veined. But some of the bodies were green, or blue, or orange, or brown, or pink. Some were patterned in tones of the same color, some were marked with black or brown, and other bodies were solidly colored. And some of those colors were poster-bright, some were muted, some powdered. But none were glistening and metallic like those of the shade-seeking damsel flies.

And their wings of gauze! Four, long, intricately matching, transparent wings on every dragonfly. On some the wings were tinted or veined, or tinted and veined, in the basic color of the body. On others, whatever the body color, the clear wings were veined in black, and many of these wings were embroidered with polka dots or with dark semaphore designs, one or more per wing.

Except for the dashing dragonflies, the gurgling creek water, and earthbound me, the noontime marsh was silent and still. I crossed the creek on two angled rocks to look into a hidden pool gouged out by the swirling waters of a bygone hurricane, and here I found a dragonfly depositing her eggs.

She was large. Her swollen abdomen was dusty-pink discreetly marked with brown, and her wings were subtly tinted and darkly veined in deep, rich chocolate. She was running in place, as it were, executing small, vertical, clockwise circles in the air above the pool.

She was facing nine o'clock, rising rearward to twelve, and dropping past two, three, and four, so that at the bottom of each circuit fully half of her abdomen was immersed in the water and swept forcibly forward as she completed the circle, rising into another revolution at the same spot and without a pause.

If she was at all aware of me, she made no sign. I stood, then sat, on the bank not more than three feet away from her, watching her activities with keenest interest. I had never known exactly how female dragonflies "washed the eggs from their abdomens" as I'd read some did, and here, before me, was a live performance.

(Not all dragonflies lay their eggs in the same manner. Some species strew their eggs into quiet waters as they fly low above the surface. Some insert each egg carefully into the underwater stem of an aquatic plant. Some, as did this pink giant, wash the eggs extruding from their abdomens into a quiet pool.)

I had been watching the pink dragonfly for probably ten minutes, during which time she had neither slackened nor paused, when I became aware that a male dragonfly in green-braided-with-black was perched on a nearby jewel weed, also watching the performance. Without warning, the green dragonfly suddenly launched himself upon the pink one, knocked her off-balance and out of rhythm, then zoomed off down the creek.

The pink female caught herself before her wings touched the water and she resumed her circling as though no interruption had occurred. The green male returned to the jewel weed and continued to stare at the female with his great black eyes.

Again, without any reason that I could imagine, he launched himself like a missile, knocked the female aside, zoomed off, then back, and knocked her aside again. Immediately, he returned to his perch on the jewel weed, and she, apparently unperturbed, resumed her circling dance.

For another ten or fifteen minutes the green male eyed the circling female and then he launched an all-out attack. He hit her from one side and then from the other, knocked her this way and threw her that. And, this time, when he zoomed off and she regained her balance, she took off in pursuit, zooming after him, up over the earth dam, across the pond, and they were lost to my sight in the regions of the upper pastures.

I waited by the pool. In five minutes or so the pink dragonfly returned. She took up her circling where she had left off, in precisely the same spot in the muddied waters of the pool.

I watched her for probably another twenty minutes. The green male did not return—I have no idea what his behavior constituted or what the incident meant, if anything—and her vertical circling continued without pause or variation.

I left her and walked over the earth dam to sit by the shallows at its western edge. Here, in the shadows cast by the pier, a dozen tiny damselflies were depositing their eggs. These damselflies were less than an inch long, slender almost to the vanishing point, with bodies of palest lavender, and clear, practically invisible wings. They laid their microscopic eggs one at a time on the undersides of algae and the leaves of the tiny duck weed, curving their tiny abdomens daintily over the margins.

Several small painted turtles pushed their gray-green-and-yellow heads into the air above the waters of the pond, spied me, and scrambled for the bottom. A nearly grown family of mallard ducklings rested in the shade of the elderberry bushes at the far end of the dam. One upstretched yellow bill angling against the low green grass betrayed their hiding place.

For many minutes I watched the egg-laying damsel flies, watched the peek-a-booing turtles, watching the markings of water currents and air currents upon the surface of the pond, watched the absolutely motionless mallards. . . .

Then, back at the hidden pool, I watched the dusty-pink dragonfly still washing away her eggs, still at her driven, vertical circling—like someone jumping an endlessly turning rope.

The green dragonfly was nowhere in sight. The pink dragonfly showed no signs of fatigue. How long could she possibly continue her egg-laying dance? I looked at my watch. Nearly two full hours had gone by. I sighed. Looked at the pink dragonfly. Looked at my watch again. Weekend guests were on their way. The naturalist reluctantly abandoned her post and the hostess went unwillingly back to her kitchen.

Towhee Talk

Several years ago, the great and the learned among American ornithologists met in solemn convention and reached an earnest conclusion. Following lengthy debate and sober voting, the assembly announced to birders in particular and to the public at large that the call note of the bird *Pipilo erythrophthalmus* is very definitely "towhee" and not "chewink" as many people would have it, and that the common name of that bird is officially "towhee."

This announcement agreed with and underlined the same decision made many years previously by an earlier congress of the American Ornithologists Union and presumably, it has settled the matter once and for all. But there is grumbling and dissatisfaction, and I suspect that the ears that heard "chewink" are still hearing "chewink," regardless of official pronouncements—and whatever it may be that the towhees are actually saying.

In my own area the towhees, and there are a goodly number of them, go around spring and fall calling "Louise!" (pause) "Louise!" (pause)—though lately I've noticed some of them have been calling "Therese!" "Therese!"—in somewhat anxious tones.

In the summer months, beginning in early June, their quick calls, not at all anxious, are simply "wink" or "whee," with an occasional short introductory sound that could be a "ch" or a "t." Never, in all my years of listening, have I heard a towhee say either *"che-*wink" or *"to-*whee." And I've listened. Hard.

On the translation of the towhees' *song,* though, there is not, to my knowledge, a breath of controversy. Throughout the Victorian era of sweetness and light the towhees sang "Sweet-bird-sinnnnnnng" although Thoreau reported the song as "Hup-hup-heeeeeeee". Ernest Thompson Seton rendered it "Chuck-burr-pilawilawila" but few people seemed to agree with him. Today it has been firmly established and, I believe, generally agreed upon that the towhee's song is a distinct, musical, unmistakable "Drink-your-teeeeeeee!" Just a simple three-note song with the last, trilled syllable higher than the rest. Usually the "your" is the lowest note, with the "drink" intermediate between them, although, sometimes, the "drink" and the "teeeeeeee" may be sung on the same level.

The eastern, rufous-sided towhee, then—making only one, although variously translated, call note and singing only one clear, distinct, unmistakeable song—is an ideal bird for a beginning birder to identify by song or call alone. Or so, once upon a time, I thought.

Not yet birder enough to recognize voices, I was thoroughly startled late one afternoon, when birding in an area several miles removed from my usual stamping grounds, to hear a distinct, musical, unmistakeable "Drrrrrrrrink-your-tea!" Some relative of the towhee, no doubt, but a new bird for my life list! I was all a-tingle. It took me just less than one hour of searching among the tangled bushes and low trees to home in on that ventriloquistic song.

There, sitting in a persimmon tree, second branch from the bottom, I found a towhee singing over and over, with intense concentration and scarcely a pause, his own version of his species' song. "Drrrrrrrrink-your-tea!" He didn't turn a feather or blink an eye when I crept beneath his tree. Just "Drrrrrrrrink-your-tea!" he solemnly admonished the irate me standing below.

This experience shook me more than a little, for, at that long-ago date, I was securely unaware of the frustrating variables that clutter and complicate the study of practically any living thing. At that time I sincerely believed that all birds of the same species sang only one song, that it was the same song, that it was sung without variations with the same quality of voice and given the same manner of delivery. And I thought, moreover, that they all sang that song instinctively and perfectly from the very first try.

But towhees taught me otherwise. Towhees taught me, first, that I must recognize the species voice rather than a rigidly sylla-bled song, else how could I ever recognize the cardinal, the song sparrow, the catbird? Towhees taught me that even a three-sylla-ble song can be adapted to quite a few variations. And towhees taught me that only with constant and repeated practice can many of them develop even passably musical voices or master the in-tricacies of their species' song. And some of them never do.

At this moment, as I write these words, a male towhee is perched on a low limb of a sassafras tree at the edge of the woods not far from my window. He is solemnly, intently and repeatedly singing his version of the towhee song. But singing is not the word for what he is doing. Over and over and over again he is spitting out three quick, short, sharp syllables. "Drink! Your! Tea!" Sometimes he pauses before the last note and I think, "This time he will trill it," but it comes out an even shorter and more explosive spit.

And note that this is no young male of this year's hatching. This fellow is more than a year old. He spent the entire month of September, last year, and part of October practicing that same short-noted, sharp-toned song.

He spent the winter here, not singing, but very much in evidence: flying low to the feeding station under the lilacs with the white outer feathers of his tail draperies trailing gracefully behind him, hopping from branch to ground with a jaunty, carefree air though the snow swirled around him on frigid winds, scratching through the snow with one foot and then the other like a slender young rooster from the henyard.

In March he began to sing again, spitting out the same unhappy version of the pleasant towhee song.

In spite of his song—surely not because of it!—a little brown female accepted him and his territory. They built their first nest under the honeysuckle at the base of a white pine and their second in the honeysuckle on the fence behind the choke cherries. I don't know how many youngsters they successfully fledged, but each nest held four white-dotted-with-reddish-brown eggs.

I had hoped the responsibilities of fatherhood might enrich his voice. But there he sits, in classic towhee pose, his black head thrown slightly back, his long tail down for balance, his white underparts and rufous sides gleaming, his red eyes soberly staring, his dark bill open wide—and from the green temple of the sassafras tree, in ritual repetition, over and over and over, three short, sharp, explosive spits!

Every summer, without fail, I hear at least one towhee having trouble with its singing, but, until this particular bird came along, I always assumed that by the time he was ready for mating the bird had mastered his song. However, I had known one that worked out a compromise solution to the problem. He started out just as this present one did, with three short, sharp, explosive syllables. After a month or so of practice he was able to hold the first two notes a little longer and to diminish their sharpness, but the third syllable still came out as an explosive spit. Eventually, and for two years, he sang a simple, strongly accented, two-note song, just "Drink-your!," leaving that third difficult trilling syllable to the birds that could handle it.

Through another summer another towhee sang only the last two syllables of the song, leaving off the "drink," slurring the "your" and trilling out the "teeeeeeee" to great lengths.

Once I witnessed something interesting, although I'm not sure just what it was. On a July evening I was picking raspberries in the orchard. From a low branch of an apple tree a towhee assiduously practiced his song. Each of the three syllables was uttered separately and sharply, with a definite pause between each syllable.

After several unimproving repetitions of this, another towhee in a sumac beyond the orchard suddenly sang out the song at a lovely pitch, trilling the "teeeeeeee" long and beautifully. Immediately the towhee in the apple tree repeated the notes exactly as he had been doing, shortly and sharply. Again the towhee in the sumac sang the musical notes and the towhee in the apple tree replied in tones sharp and unlovely. This singing lesson, if, indeed, it was a lesson, continued all the evening (and was still going on when darkness and mosquitoes sent me scurrying into the house) without the slightest improvement, that I could hear, in the song of the towhee in the apple tree.

There are other birds, of other species, I am told, who also must learn their species' songs, but none of them seem to have so much trouble or to afford their listening public so much amusement as these towhees, struggling with their three distinct, musical, and unmistakeable notes that have to be sung exactly right.

Everybody Knows Katydid

Six-weeks-'til-frost!" "Six-weeks-'til-frost!" that's the real message of the first Katydid calls, an old man in the Appalachian mountains told me. "Never fails!" he said, and I'm sure he is right, but I forgot to ask him when the Katydids begin to sing in his high hill country.

A few weeks earlier than hereabouts, I should think, for our first killing frost arrives in late September or early October and approximately, if not precisely, fulfills the prophecies of the Katydids who begin their abrasive music in the darkness of mid-to late-August nights.

The frosts bring a definite end to the rasping concerts of the Katydids; and if a frost in early September catches us and our gardens by surprise, it makes the Katydids pay, with an early death, for their inaccurate and faulty predictions.

But, what, exactly, is a Katydid? A British entomologist complains that American authors who write of the rural scene almost invariably mention Katydids, but never tell what a Katydid is. "Every American reader," says this British scientist, "knows so well what a Katydid is that it is unnecessary for the writer to explain. . . ."

Once I would have thought his statement true, but in recent years I have had so many letters asking about Katydids that I suspect many American readers have only a vague idea of what a Katydid may be.

Since the Katydid seldom has anything to do with the plot of the book, it doesn't matter too much, I suppose, whether the reader knows or not, except that all Katydids are exclusively North American, and the true Katydids, the ones that speak English, are resident only in our southern and eastern states. So it seems that out of plain patriotism, we ought to be better acquainted.

True Katydids belong to the family of the long-horned grass-hoppers. Like most Katydids they are green—leaf-green. Their green bodies are about two inches long and their wide green wing covers are often longer than their bodies. They have six green legs, with the "knees" of the rear pair folded high on either side, for this rear pair are extra-long, grasshopper-type jumping legs. Their green antennae, their "horns," are much longer than their bodies, and they carry them, most of the time, arched over their backs in graceful, sweeping curves. Only rarely are they turned forward, for their use as sensory organs.

The bodies of Katydids are laterally compressed. That is, their bodies look as though they have been pressed between two boards, from both sides, so that they are thin and high, and their wings fold against their sides rather than across their extremely narrow backs.

Their leaf-greenness makes them almost invisible in the bushes and the high tree-tops where they spend their lives dining in vegetarian abundance and with scarcely any effort on the lush green leaves that surround them.

And from among these leaves they sing. The sounding aisles of the night woods ring for six late-summer weeks with the deaf-ening resonance of their scratchy songs, which aren't songs, ex-actly, in spite of the verbal "Katy-did!" "Katy-didn't!"

Fiddle music is what they make, and it is produced by the scraping of green wing against green wing over and over and over again. "Stridulating" is the word for it.

On the underside of the left wing the true Katydid has a file with about fifty-five teeth. On the upper side of the right wing is a single raised ridge called a "scraper." The Katydid lifts its wings —left wing always on top—and by moving them rapidly back and forth in a scissors-like motion it draws file and scraper across one another and produces its extraordinarily loud "Katy-did!" "Katy-didn't!" argument.

Each rasping of the wings is said to produce only one syllable of the call, so that the shortened "She-did!" is simply a rasping in-out; the "Ka-ty-did" is in-out-in; and "Ka-ty-did-n't!" is in-out-in-out.

A scientist whose business it is to figure such things once calculated that a Katydid, in the course of its six-week season of stridulating, rasps its wings together some fifty-million times.

How can this simple rasping together of a chitin scraper and a file, even on several hundreds of insect wings, raise such a rackety din in the night time woods? How can the solo rasping of the lone Katydid that starts the evening's concert be so remarkably loud?

It's all done with mirrors! On each wing, directly adjacent to the scraper and to the file is a flat, D-shaped area of clear chitin called a "mirror." This is a resonance area, much like the head of a drum, that transfers and amplifies the gratings of file and scraper —plus the accompanying vibrations of the wings—to the surrounding air.

Only the males are equipped with the noise-making apparatus. The females maintain a discreet silence—perhaps they are all named Katy—although they can make a faint complaint or two if they are disturbed or inconvenienced.

But the females have ears to hear the stridulated fiddling of the males. Their ears, and the ears of the males, too, are located on the tibia of the forelegs, just below the "knee" joint. We do not know what the range of their hearing may be or whether they hear anything beyond the calls of their own kind and their own making.

The probable reason for the Katydids' treetop concerts' strident clamor is that there just isn't, ever, any very great abundance of true Katydids and the clan has to be called together, and kept together, for purposes of courtship and mating.

Katydid behavior may vary from one locality to another. In my area the earliest soloist does not start its harsh fiddling until true darkness has fallen. And night time is here, beyond question, when the full body of the treetop musicians swings into concert. For three or four or five hours, then, with never a pause, they rock the aisles of my small forest with the intensity of their ringing, rasping, grating din.

The night is still dark when they stop, dark without the slightest hint of dawning, and they stop, suddenly and completely. Once in a while a soloist may pick up here or there with a measure or two, but it doesn't go on.

They don't start at the same hour nor stop at the same hour even on consecutive nights when darkness and temperature and humidity would seem to be closely the same. As with the flashing of the fireflies, it must be some delicate balance of these three elements that triggers the fiddle-playing of the Katydids—and some imbalance that stops it.

These earsplitting woodland extravaganzas go on every night (unless the night is wet and cold) for just about exactly six August and September weeks. And then they stop. Usually they stop because the musicians are killed by the heavy frost their music has been predicting; but even if the frost fails to arrive on schedule the concerts lose their spontaneity and their enthusiasm, become desultory, cease altogether. The season is over, the performers are exhausted, they creep among the leaves without interest, and they die because it is time to bow off the stage.

At any rate, the apparent purpose of the six weeks' gathering is to provide for the continuation of the species. Once mating has been accomplished, the female, with her green sickle-shaped ovipositor, places 100 to 150 flat, slate-colored eggs, one at a time, in the ridges of tree bark or in the soft stems of shrubs. (It is the angle-winged Katydid that lays her eggs in neat, overlapping trim on the edges of leaves and twigs.)

When spring arrives, the developing bodies within the flat eggs push them into more rounded shapes, and in early May, tiny Katydids with bodies only an eighth of an inch long (their antennae and long jumping legs, stretched out fore and aft, give them an overall length of half an inch) push out of the eggs, casting their first skins as they hatch.

They are delicate things, rarely to be glimpsed as they learn to manage their six legs, take dainty leaps, and try out their miniscule jaws on tender vegetation.

They shed their skins three more times in the next seven or eight weeks and by the end of July they are nearly adult and are packing their incipient wings in thick parcels on their backs.

Sometime in early August the fifth skins split and, after heroic struggles, the newly winged, fully adult true Katydids step forth and begin a slow progress toward the tops of the woodland oak trees and maples, and of the apple trees and the cherry trees in the orchard.

And from those leafy vantage points, some time soon, the males of this generation of true Katydids will play, for their first time ever, a ringing, raucous, grating concert that will be exactly the same as the one last year and the year before and the year before that—"Katy-did!" "Katy-didn't!" "Katy-did!" "She-did!" "She-did!" "She-didn't!"

Crickets, Crickets, Crickets

merican crickets come in several assorted varieties, but in most parts of the United States when we say "crickets" we mean the shiny black-and-dark-brown field crickets that live under stones and boards and piles of vegetation or in burrows they dig for themselves in the earth—the same big crickets that get into our houses in the fall and startle or surprise us with the sudden loud ringing of their chirping calls within the closed walls of our dwellings.

There are some true house crickets in the United States—brought in purposely or inadvertently, from Europe, by colonists and later arrivals—but they are not particularly plentiful here and they are not *the* cricket.

When we mean house cricket or tree cricket or bush cricket or mole cricket or cave cricket we say so. When we mean field cricket we simply say "cricket."

These crickets, along with fireflies and butterflies are probably the most "socially acceptable" insects around. But, just as fireflies are known mainly as flashing lights and butterflies as flutterings of lovely wings, so crickets are known as tinkling chirpings that provide an almost unnoticed musical undergirding for our summer-into-late-Autumn days and nights.

Like all other living things, crickets are something more than the sum of their parts and much, much more than the mechanical music boxes we unthinkingly assume them to be.

They are sleek, dapper, dark-colored, living animals with a measure of adventure and of unpredictability in their individual lives. Physically far more bouncy and ebullient than their close relatives, the true Katydids, crickets are also, to a small degree, less tightly bound by species restrictions.

Whether or not their shape has anything to do with it, Katydid bodies are flattened from side to side but cricket bodies are flattened, less radically, from top to bottom—dorso-ventrally flattened. Their glistening dark wings lie flat on their backs with the edges curved downard to cover their rounded sides, and their antennae, only half as long as their bodies, explore the world before them.

Their shiny-clean bodies are fully an inch long and their powerfully muscled, rear jumping-legs are somewhat longer than their bodies. Both sexes have a pair of spines, their "tail feathers," extending at an angle from either side of the rear of the abdomen, but the female may be distinguished from the male by the very long, stiff ovipositor that protrudes between her tail feathers from the bottom-rear of her abdomen.

Crickets will eat almost anything. They eat fruit and grain and green vegetation; they eat small insects, other crickets, and any carrion that they find. In our houses they will eat all sorts of foodstuffs—breadcrumbs, lettuce, meat, sugar, and will nibble at rubber balls and at rugs and clothing made of natural fibers.

Cricket concerts are almost as varied as their menus. They are not restricted to playing during only certain hours of darkness. They play, literally, when and as they feel like it.

Their grand concert begins in the late afternoon, grows louder as early darkness settles down, continues more or less strongly through the night, swells into a great outpouring just before, or at, dawning, and dwindles into small-group-playing or into solo parts through mornings, noons, and early afternoons.

As long as the temperature doesn't drop below fifty-five degrees Fahrenheit in its hidey-hole, some member of the cricket orchestra is playing from sometime in July or August until nearly snowfall.

And it is a cricket orchestra. Crickets stridulate. They are fiddle virtuosos. But only the males are musicians for only the males have wings fitted out as musical instruments. Only the males have wings with clear, resonant areas for musical sounding boards.

On the under side of each male cricket wing is a file and on the upper side of each male cricket wing is a scraper. The male cricket lifts its wings at approximately a forty-five degree angle and, using the same scissors-like motion as the Katydid uses, it brushes file across scraper and produces its own individual music, at its own individual time, in its own individual way.

Crickets are not restricted to playing always with the same

wing. With their convenient arrangement of files and scrapers they can switch wings any time they choose, although, in practical use, it seems, they almost always play with the right wing over the left.

Crickets are free to play at any hour, or through all the hours, of the day or the night, so long as the air temperature is warm enough to keep their small bodies functioning.

And crickets are not confined to a scant six-week season of late-summer-into-fall music making. This year's generation of crickets may begin its concerts two or three weeks ahead of this year's Katydids, and the crickets are still going strong in their covert niches long after killing frosts have done away with the Katydids in the tree tops and the bushes.

Crickets fiddle, albeit somewhat rustily, on warm, sunny days in November and even in December. If they can manage to live through the winter, and many of them do, they thinly welcome the spring in March, and fiddle sparingly through April and into May.

How long they may live I do not know, though I feel reasonably certain none live through two winters. I had been sure the winter veterans did not live into a second summer, but this year, through the entire month of June when this season's elfin youngsters were scarcely three-eighths of an inch long, I heard constant rich and full-toned cricket solos coming from among the blossoming chives and thyme at my kitchen door.

During July, while the youngsters grew prodigiously and some of them sported wing buds—I saw none that had reached full musician status—full-blown cricket solos, duets, and trios were being played right there among the flagstones.

I never located the players so I do not know whether they were extra-hardy oldsters left over from last year's orchestra or highly precocious youngsters just hatched this spring and eagerly trying out their musical wings.

Springtime is hatching time for crickets. Flea-sized and energetic they bounce out of the pale-yellow, banana-shaped eggs their mothers inserted, one by one, into the earth last fall. Shedding their first skins as they hatch, they move out in tiny cricket leaps to meet the great world on tiny cricket terms.

By May they are quarter-inch-long leaping brownies, getting into baskets, boxes, flower pots, houses, under foot, under broom, under mower, under tire, as well as inside birds, and skunks, and shrews, and mice, and other small and hungry animals.

The survivors shed their small skins from seven to, perhaps, a dozen times before they emerge in late July with wing buds in two thick packs upon their backs. With the following, final, molt they emerge fully equipped with strong shining wings that will never fly but with which the males of their generation will play the tinklingest music in the entire insect world.

The Rollicking
Carolinas

A busy twittering among the brambles on the edge of the woods, puffs of snow flying from froth-heaped branches, and two small birds dashing from the tangle. . . .

They checked their flight, these birds, and perched side by side, with bare toes curled through an inch of cold rime onto the black branch of a dogwood, and they looked me over with bright brown eyes neither welcoming nor afraid.

I stood ankle-deep in dry leaves topped with snow and looked them over with eyes less brown than theirs but bright with the warmest of welcomes, for just one glance took in the red-brown backs, the buffy flanks, the white eye-stripes, the impudent, unmistakable cock of their tails. Carolina wrens! Their species has been missing from my woods for several seasons now and my ears have ached with the longing for their non-existent voices.

As though he read my thought, the male dropped his tail, raised his head, and, in as solemn a manner as a Carolina wren ever does anything, he loudly announced to the disinterested woods, "Miz Leis'ter, Miz Leis'ter, Miz Leis'ter, Miz."

"Why, how do you know, you little stranger?" I asked aloud.

The birds, startled at the sound of my voice, flicked their heads, flicked their tails, flicked their wings, and they were gone —to seek out some other woodland thicket for their habitation. For I have not heard them since, and my woods and my life are distinctly the poorer.

Not that Carolina wrens are old, old friends of mine. I met the chunky little fellows only seven or eight years ago when a pair of them invaded the garage, thrown open for an April weekend of lawn and garden duties.

They didn't really invade the open garage, they frolicked in. On a jolly binge of exploration, they exulted to one another over every shelf, over every cranny, over every concave surface. Standing atop the car, bodies stretched to the utmost, bright eyes peering at the multitude of exciting possibilities, they made the entire holding ring with their merriment. Finding a nesting site was no solemn obligation or worrisome responsibility for this rollicking pair. They had a ball.

Then came human intervention. Even before all the crannies were looked into or all the niches investigated, the garage doors were closed, the adventuring pair was hurried out through an open window, and the window, too, was closed behind them.

Nothing daunted, the two frequented the raspberry tangles in the nearby woodland thickets, and the male filled the air with the effervescence of his rollicking songs while the female added her tinkling rattles and I rejoiced that they had found a close-by woodland cavity for their nest.

One morning, as I left for a day of shopping, I inadvertently closed the door to the small garden house normally left open, so that dog or cat might find shelter from sun's heat or from a shower. When I returned several hours later, I found the female Carolina wren frantically trying to break through the glass of the garden-house door and the male hysterically singing from the nearby spruce.

Curious, I opened the door, and the frenzied female flew instantly to the highest shelf, up under the roof, and settled herself into a flowerpot, out of sight except for her eyes and her bill.

From that day onward the little female, who formerly must have sat quietly and trustingly through many disturbances, dashed out over my head every time I stepped through the door. I moved the most frequently used garden tools to the garage and avoided going near the garden house except in the direst of need, but I worried that the little hen's two weeks of nesting might be in vain because of those hours she had been shut out from her five brown-sprinkled eggs.

But my anxiety was unnecessary, for several days later I noticed that both wrens were busily carrying captured insects into the garden house. And when I climbed up to investigate that potful of leaves and dried grasses, I found five baby wrens sprawled in a heap down in the hollowed center, so softly lined with white dog-hairs.

The industrious pair of wrens fledged their five youngsters, singing all the while, then moved somewhere nearby for their second nesting of the season—and also for the third. In fact, for their own reasons, not disclosed to me, they never again nested about the buildings. But they and other pairs of their species so built up the population in the general area that through the fall and early winter there were Carolinas all over the countryside— with five distinct pairs living on the fringes of my woodland and in my orchard—and their loud, rollicking, year-around songs were an exhilaration to me.

Usually their songs were three-syllable phrases repeated three or half a dozen times in quick succession, although sometimes the phrases were only two syllables long. But, whatever their length, they were extremely variable in their sounds and I found myself endlessly translating them into English. "Wheat-eater, wheat-eater, wheat-eater, wheat!" "Sweetheart, sweetheart, sweetheart, sweet!" "Per-tatie, Per-tatie, Per-tatie. Pert!" "Te-wee, tewee, te-wee, te-wee."

But there was one song they sang, though not often enough, that was not divided into these syllables. It was so lyrical, so sweet, so different, that I could not believe it was a wren making the music until I had thrice seen the lowered tail, the open bill, and heard the lilting song coming from the spot where the singing wren sat.

Through the years, with the spreading population, the songs increased, for one bird incited another; and only in the depths of the late-summer moulting season were the voices of the Carolina wrens ever even nominally stilled. Even in a storm of blowing snow some wren in some protected tangle always found life worth singing about.

But the bitter winter of '76 passed into the bleaker winter of '77 and the wren songs in my woodland grew fewer and fewer, tapered off, disappeared altogether. I don't know when the last song was sung, because, of course, I didn't know it was the last. But there came a day when I realized that my ears were straining to hear the excitement of a wren's song and that there was none, and that there had been none for quite some days. And that I had not seen a jaunty, rusty fellow for a long, long time.

Even then I supposed they'd only gone off to deeper shelter than my woods afforded, or perhaps had migrated southward from the bitter winter weather, but when spring came, and then summer, and no loud songs of the Carolina wren were superimposed on the rolling choruses of cardinal and catbird, thrasher, sparrow, oriole, robin and the whole host of singing birds, I knew that tragedy had emptied my woods of wrens while I was unaware.

I have been told, by eminent ornithologists, that as long as a bird has plenty to eat it is not bothered by the cold. Carolina wrens, as far as is known, eat mostly insects, plus a few selected berries and seeds. Some people report Carolina wrens eating suet and peanut butter at their feeders. I did not serve peanut butter, but I did have suet hanging in net bags from several tree limbs on the border of backyard and woods. Yet I never saw a Carolina wren eating there or even approaching a bag through that or any other winter, although they came often into the general environs.

But, for most of every day, I am not watching the activity in my backyard, and perhaps they did eat my suet unobserved by me. And perhaps, over a long period of privation, the suet was not enough. Perhaps it lacks vitamins and minerals necessary to the life of a Carolina wren accustomed to the full bodies of insects. Perhaps.

But, also, I am told, the trend is cyclic. When food is plentiful and most nestings are successful, the young birds of each successive generation must wander farther and farther north to find unclaimed land where they can stake out territories of their own. If the winters in these new lands stay mild for a few years, as had happened here, large colonies of Carolina wrens develop and become accepted as part of the resident biota.

Then comes a bitter winter with severe storms and constant, unyielding cold, and the merry but ill-suited wrens are wiped out.

But they never give up. Population pressures in the home grounds will send them wandering northward again and, one year, they will be singing once more in your woods and in mine. And perhaps, just perhaps, some day, one generation will successfully adapt.

Groundhog Roles

Short, unpopular February is credited with Valentine's Day, the birthdays of two famous presidents, the worst weather of the year, Candlemas Day and groundhog day—Candlemas and groundhog day, of course, falling together on February 2nd.

If the authorities in charge of such matters had settled groundhog day two weeks later to share Valentine's Day, or, better, three weeks later to share Washington's actual birthday, the appearance of a wild, free, newly awakened groundhog at the celebration would have been far more likely and the prognostication of the weather would, I'm sure, have been equally precise.

As a matter of fact, the original European myth for Candlemas holds either a bear or a badger (usually a badger) responsible for forecasting the nature of the weather for the closing weeks of winter. How that role got transferred to the groundhog in this country, I do not know.

However, the groundhog takes his duties as a weather prophet very lightly, or not at all, and remains undisturbed in his comatose state for another fortnight or so, his emergence being more a matter of his own metabolism than a response to human whimsy.

In most of his range the only groundhog likely to be out and about as early as February second is one who failed to fatten properly during the gluttonous feasting days of the previous August and September. He is awake now only because he is hungry, and he will probably stay hungry until new green growth begins in clovers and grasses, for there is little nourishment remaining in the yellowed, leached-out herbage of winter's end.

In all my years of groundhog-watching, the earliest date on which I have ever seen a self-satisfied, adventure-on-his-mind groundhog at large was February 9th of a year when spring arrived in mid-January, and the February daytime temperatures held

steadily in the upper sixties. By February fourteenth, of that year, every groundhog in the neighborhood was astir and a dozen romances were well under way. But that was an unusual year.

Ordinarily, in my neighborhood—and it has proved true year after year after year—February 22nd is the date when most of the groundhogs end their winter sleep, clean out their burrows and start afresh into a brand-new year.

Or perhaps they awaken on the 20th or the 21st and start out on the 22nd, for it is believed that, in the wild, their natural full awakening from the almost-death of hibernation requires from twelve to twenty hours of blinking, stretching, breathing. . . .

Once, precisely on February 22nd, I saw a young male start out on his new year's adventures. He did not sit blinking in his doorway as I had expected, but, after a brief, precautionary, eyes-half-open look around, he half-waddled and half-slid down the muddy path from his woodland knoll.

His hair stood out in every direction, pressed into wild disarray by his months of sleep, and his shaggy coat was smeared with mud from frowzled head to raggedy tail.

Once clear of the woods and out on level ground he paused and blinked his little piggy eyes in the brilliance of the afternoon sunshine. He lifted his chin. He sniffed the air. He wagged his tail. He stretched his lips back and back in what could only be a smile and looked all about him with the most beatific (or was it fatuous?) expression on his face.

Still blinking his eyes, he waddled on another dozen steps, stopped, lifted his chin, sniffed the air, wagged his tail, and bestowed, once more, his smiling benediction on the world. This blessing he repeated over and over again until he disappeared from my sight into the weed-field beyond the orchard.

Two days later I met him near the entrance of his den. He was sleek and clean, no longer blinking in the light and, while he was not hanging heavy with fat as he had been in the fall, neither was he thin.

Groundhogs do use their fall-stored fat to get them through the winter, but they don't use much of it during hibernation. Very little energy is required by a body when its temperature is reduced to around forty or fifty degrees Fahrenheit, its circulation is almost at a standstill, and its need for breath reduced to an intake only once in every six minutes.

It is a strange paradox that groundhogs go into hibernation, usually in October, while their food supply is still rich and plentiful, and come back into the active world during February and March when, quite often, snow is still deep upon the ground, and when, even more often, there is little food to be found.

It is in this period between their awakening from winter sleep and the appearance of new green growth among the clovers that groundhogs really use up their store of accumulated fat, sometimes several pounds in one month.

But, then, this first month or six weeks after awakening is romance time among the groundhogs, and the males, in particular, use up their reserves in seeking out the dwelling places of their lady loves. The females, reportedly, stay quietly in their burrows awaiting their chance-sent suitors, and the groundhogs we see abroad upon the face of the earth at this time are all brave knights in shining armor.

Each of these knights seeks until he finds the burrow of an unmated female and, if the lady is willing, he moves in with her for the duration of the mating season. All groundhog courting takes place underground, and the rule of monogamy, at least for each season, is a convention seldom spurned.

But, just before the young are born, the male, probably at her insistence, moves out of the female's burrow. He returns to his own homestead and spends the summer looking out for himself while his erstwhile mate rears their three to five, usually four, youngsters all on her own.

The young male groundhog whose emergence into the after-winter world I so fortunately witnessed apparently had difficulty locating a mate for himself that spring, for he kept returning to his own burrow on the woods-knoll for quite some time.

One sunny afternoon in early March I saw him leave his burrow, go down to the marsh, cross the creek, and climb the pasture hill toward a complex of groundhog holes around an outcropping of rock. I knew there was a young female in residence up there, but I was also fairly certain that a big grizzled male had already claimed her, so I followed him at a distance and watched the proceedings with a great deal of interest.

The young female happened to be topside nibbling on some sprouting green by the end of a rock and she saw the young male approaching. She ran at him with a thin "You-get-out-of-here" shriek, but the young male refused to leave. She pushed him. He pushed her back. She ran away from him. Ignored him. Turned her back upon him and resumed her interrupted nibbling.

But the young male would not be rebuffed. He pressed close up against her. Pushed her from one side. Pushed her from the other. And it soon became evident that he was trying to push her toward the entrance to her burrow.

She resisted, evaded, doubled back around the rock. He pursued, persisted, ambushed her by short-cutting across the low end of the granite. For a little while there was a great deal of grunting and squealing, of pushing and wrestling and running around the rocks, and sometimes it was difficult to know which of the young groundhogs was which. But after a while the young female broke away from her insistent suitor and plunged down the precipitous entrance into her burrow, with the young male chasing close behind her.

Pandemonium broke out then, down there in the depths of the groundhog's dark tunnel. There were snarls and shrieks and growls and thumps, and suddenly the young male erupted from the burrow entrance with the old grizzled male hanging to him, long incisors buried in the flesh of the young male's buttocks.

Once out in the open, free of the narrow burrow walls, the young male shook off his assailant and fled, squealing and bloody, down the hillside with the enraged householder still nipping at his flanks.

Down the hill, across the flat, under the fence, and into the marsh, and there the grizzled male ended his pursuit—but the young male did not know it. Still fleeing and still squealing, he dashed full-tilt through the reeds and the burgeoning skunk cabbage and fell headlong over the high bank into the cold waters of the creek—to the consternation of the lone Guernsey cow drinking there, under a willow, on the low and bankless side.

The cow threw up her head and backed away from the thrashing water. She stood there staring, allowing her mouthful of water to drizzle away to the ground and looking as though, if she'd had eyebrows, they'd have been up around her horns somewhere.

I'm afraid I laughed aloud, and at length, at this hilariously unexpected bit of slapstick. The guilty young male's surprised landing in the cold waters of the creek was a pat enough ending to the story, but the Guernsey cow's being right at that spot at that particular moment—and her Imogene Coco reaction—was a scenarist's touch I could never have imagined.

However, the Guernsey cow and I did regain our composure, and both of us watched in thoughtful silence as the deflated Romeo, dripping wet and bleeding from his wounds, limped slowly along the edge of the woods, hobbled up the shadowed knoll, and disappeared down the sloping entrance into his own dark place of refuge.

Designed by Barbara Holdridge
Composed by Com Com, Allentown, Pennsylvania
in Palatino with display initials in Torino
Printed by Haddon Craftsmen, Bloomsburg, Pennsylvania
on 60 lb. Old Forge Opaque paper
Bound by Haddon Craftsmen, Scranton, Pennsylvania
in Kingston Natural Finish cloth with Multicolor Antique Thistle endpapers
Jackets printed by Rugby, Inc., Knoxville, Tennessee